GOVERNORS STATE UNIVERSITY LIBRARY
W9-AHK-999
3 1611 00352 7683

DEMCO

Resources for the Study of Real Analysis

GOVERNORS STATE UNIVERSITY
UNIVERSITY PARK, IL

QA
301
.B73
2004

© 2004 by
The Mathematical Association of America (Incorporated)
Library of Congress Catalog Control Number 2004102964

ISBN 0-88385-737-5

Printed in the United States of America

Current Printing (last digit):
10 9 8 7 6 5 4 3 2 1

Resources for the Study of Real Analysis

by

Robert L. Brabenec
Wheaton College

GOVERNORS STATE UNIVERSITY
UNIVERSITY PARK, IL

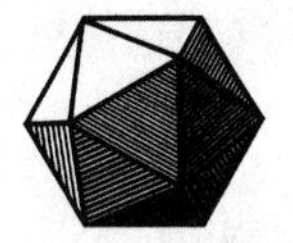

Published and Distributed by
THE MATHEMATICAL ASSOCIATION OF AMERICA

CLASSROOM RESOURCE MATERIALS

Classroom Resource Materials is intended to provide supplementary classroom material for students—laboratory exercises, projects, historical information, textbooks with unusual approaches for presenting mathematical ideas, career information, etc.

Council on Publications
Roger Nelsen, *Chair*

Zaven A. Karian, *Editor*

William Bauldry, Daniel E. Kullman
Gerald Bryce, Stephen B Maurer
Sheldon P. Gordon, Douglas Meade
William J. Higgins, Judith A. Palagallo
Mic Jackson, Wayne Roberts
Paul Knopp

101 Careers in Mathematics, edited by Andrew Sterrett
Archimedes: What Did He Do Besides Cry Eureka?, Sherman Stein
Calculus Mysteries and Thrillers, R. Grant Woods
Combinatorics: A Problem Oriented Approach, Daniel A. Marcus
Conjecture and Proof, Miklós Laczkovich
A Course in Mathematical Modeling, Douglas Mooney and Randall Swift
Cryptological Mathematics, Robert Edward Lewand
Elementary Mathematical Models, Dan Kalman
Environmental Mathematics in the Classroom, edited by B. A. Fusaro and P. C. Kenschaft
Essentials of Mathematics: Introduction to Theory, Proof, and the Professional Culture, Margie Hale
Exploratory Examples for Real Analysis, Joanne E. Snow and Kirk E. Weller
Geometry from Africa: Mathematical and Educational Explorations, Paulus Gerdes
Identification Numbers and Check Digit Schemes, Joseph Kirtland
Interdisciplinary Lively Application Projects, edited by Chris Arney
Inverse Problems: Activities for Undergraduates, C. W. Groetsch
Laboratory Experiences in Group Theory, Ellen Maycock Parker
Learn from the Masters, Frank Swetz, John Fauvel, Otto Bekken, Bengt Johansson, and Victor Katz
Math through the Ages: A Gentle History for Teachers and Others (Expanded Edition), William P. Berlinghoff and Fernando Q. Gouvêa
Mathematical Evolutions, edited by Abe Shenitzer and John Stillwell
Mathematical Modeling in the Environment, Charles Hadlock
Mathematics for Business Decisions Part 1: Probability and Simulation (electronic textbook), Richard B. Thompson and Christopher G. Lamoureux
Mathematics for Business Decisions Part 2: Calculus and Optimization (electronic textbook), Richard B. Thompson and Christopher G. Lamoureux
Ordinary Differential Equations: A Brief Eclectic Tour, David A. Sánchez
Oval Track and Other Permutation Puzzles, Robert B. Ash
A Primer of Abstract Mathematics, Robert B. Ash
Proofs Without Words, Roger B. Nelsen
Proofs Without Words II, Roger B. Nelsen

A Radical Approach to Real Analysis, David M. Bressoud
Resources for the Study of Real Analysis, Robert L. Brabenec
She Does Math!, edited by Marla Parker
Solve This: Math Activities for Students and Clubs, James S. Tanton
Student Manual for Mathematics for Business Decisions Part 1: Probability and Simulation, David Williamson, Marilou Mendel, Julie Tarr, and Deborah Yoklic
Student Manual for Mathematics for Business Decisions Part 2: Calculus and Optimization, David Williamson, Marilou Mendel, Julie Tarr, and Deborah Yoklic
Teaching Statistics Using Baseball, Jim Albert
Writing Projects for Mathematics Courses: Crushed Clowns, Cars, and Coffee to Go, Annalisa Crannell, Gavin LaRose, Thomas Ratliff, and Elyn Rykken

MAA Service Center
P. O. Box 91112
Washington, DC 20090-1112
1-800-331-1MAA
FAX: 1-301-206-9789
www.maa.org

This book is dedicated to the hundreds of Wheaton College students who have taken my course in real analysis during the past forty years. Their enthusiasm and contributions enriched these classes and helped to shape the material in this book.

Contents

Section 3. Enrichment Problems

III Essays 125

Section 1. History and Biography

Section 2. New Looks at Calculus Content

Section 3. General Topics for Analysis

Preface

This book is a collection of materials I have gathered while teaching a real analysis course every year for more than thirty-five years. I prepared it with the hope it will benefit and enrich the experience both of students who take a real analysis course, as well as those who teach it. This collection is intended to supplement a traditional real analysis textbook, where teachers and students may choose items of interest to them. It is my conviction that such supplementary materials have a much greater chance of being used in an analysis course if they are readily available in one place.

Part I contains materials that provide the greatest benefit if read before the real analysis course begins. Because such a course assumes knowledge of topics from calculus of one real variable, review of this material in advance frees the student to concentrate on new content and theoretical emphases in analysis without having to revisit the calculus at the same time. The outline of a traditional calculus course allows a student to check necessary topics for review, whereas the calculus review problems offer the opportunity to refresh necessary skills which may have lain dormant for some time. Many hints are provided with the problems to encourage students to carry out this review.

The problems in Part II are intended to supplement the ones usually found in an analysis text. There is a wide variety of problem types, and each one has exercises for the student to attempt. Most of them contain explanatory detail about the historical background of the topic or how it fits with other parts of analysis. I try to present a topic from a variety of perspectives in order to enhance learning and understanding.

The entries in Part III are called essays, and represent short discussions of a particular topic from calculus or analysis. Some of these are content-oriented, whereas others are intended to give an alternate perspective after the student has first studied the material in a traditional manner. The collection of six essays based on time periods presents a historical overview of the development of analysis by concentrating on the main individuals who were influential in this development.

Part IV contains a collection of five supplementary readings to illustrate the variety of materials that are available beyond a standard textbook. The annotated bibliography contains many references that can be read with profit by faculty and students. There is information for each one to help readers decide which might best fit their interests and needs. We are fortunate to be living at a time when there is a growing interest in these kinds of supplementary materials—the past twenty years especially have seen many new entries of a historical, biographical, or expository nature.

Let me explain how I use these materials in my analysis course. I encourage students to work on the calculus review problems before the analysis class begins. Since our course is taught in the fall semester, I give them the problems before they leave for summer vacation. At the beginning of the course, while presenting the abstract material on properties of real numbers and sequences, I use Essays 7 and 8 to review the material on derivatives and series from calculus. This not only gives students a more familiar alternative to the abstract material, but it also provides an introduction to the spirit of careful organization and attention to detail that is essential for work in analysis. Students read one of the six historical essays a week and we spend some time in class discussion on the material. I use Essays 9 and 10 in the middle of the course to provide an alternative perspective for the topics of proofs and topology, and assign supplementary problems from Part II at appropriate places in the course. The material in chart form from Problems 8, 9, and 20 works well in class discussions. I like to assign some of the enrichment problems in Section 3 to teams of students and have them learn the material, write up solutions to the exercises, and make oral presentions to the rest of the class. Students enjoy this opportunity to make presentations.

Acknowledgements. I would first like to express appreciation to my students who have used different versions of these materials in classes and contributed their suggestions for improvements. I am also grateful to Bill Dunham from Muhlenberg College and Judy Palagallo from the University of Akron for reading various portions of this manuscript and offering valuable advice for improvement. My mathematics colleague at Wheaton, Terry Perciante, helped me prepare the figures for the book, and my computer science colleague, John Hayward, guided me through and around the pitfalls of working with a computer. Much of the work was accomplished during a sabbatical semester in 2002 at Cambridge University. My thanks go to Wheaton College for providing this sabbatical opportunity, to the Aldeen Fund for financial support with travel expenses, and to the Department of Pure Mathematics and Mathematical Statistics at Cambridge for their support during my time there. I am also grateful to my wife Bonnie for her support and good advice during my work on this project.

Many of you will undoubtedly be aware of additional problems, topics, or bibliographic items that can be added should there be a revision of this book. I would be glad to receive these at Robert.L.Brabenec@wheaton.edu.

I

Review of Calculus

An Outline of a Traditional Course in Calculus

Section I. Using the Limit Concept to Represent and Study the Major Concepts of Calculus for a Function of One Variable

1. How to Find Limits
   (a) Factor and cancel (when form is $\frac{0}{0}$)
   (b) Use of a conjugate factor
   (c) Divide by highest power of x (when taking $\lim_{x\to\infty}$ of a rational function)
   (d) Squeeze theorem
   (e) Know that $\frac{1}{0}$ form means $\pm\infty$
   (f) L'Hospital's rule (for 7 indeterminate forms)
2. How to Use Limits
   (a) To define continuity

$$\lim_{x\to c} f(x) = f(c)$$

   (b) To define the derivative

$$f'(x) = \lim_{h\to 0} \frac{f(x+h) - f(x)}{h}$$

   (c) To determine types of discontinuities (removable, infinite, jump, oscillating)
   (d) To find asymptotes (vertical, horizontal, slant)
   (e) To evaluate an approximating sum for area of the form

$$\lim_{n\to\infty} \sum_{i=1}^{n} F(a + i\Delta x)\Delta x$$

(f) To define the definite integral

$$\int_a^b f(x)\,dx = \lim_{|P|\to 0} \sum_{i=1}^{n} f(c_i)\,\Delta x_i$$

(g) To evaluate improper integrals

$$\int_a^\infty f(x)\,dx = \lim_{t\to\infty} \int_a^t f(x)\,dx$$

(h) To find the value for a convergent sequence

$$\lim_{n\to\infty} a_n = A$$

3. How to Find Derivatives
   (a) By the definition
   (b) By standard formulas—power rule, product rule, quotient rule, chain rule
   (c) Memorize formulas of standard functions and combine them with the above rules
   (d) Method of implicit differentiation
   (e) Method of logarithmic differentiation
   (f) Formula for inverse functions

$$\frac{d}{dx} f^{-1}(x) = \frac{1}{\frac{d}{dy} f(y)}$$

   (g) Formula for a function defined by an integral

$$\frac{d}{dx} \int_a^x f(t)\,dt = f(x)$$

4. How to Use Derivatives
   (a) To find tangent and normal lines to curves
   (b) To find velocity and acceleration
   (c) To find relative maximum and minimum points, inflection points, intervals of concavity
   (d) To make an accurate graph for a function
   (e) To solve max-min word problems
   (f) To set up and solve related rate problems
   (g) To solve differential equations (such as the one for exponential growth)
   (h) To understand the algebraic and geometric meanings of the mean-value theorem
   (i) To find approximations using differentials or Newton's method
   (j) To find limits of indeterminate forms using L'Hospital's rule
   (k) To find polynomials that approximate a given curve

5. How to Find Integrals
   (a) By the integral definition when the partition is into n equal parts

$$\lim_{n\to\infty}\sum_{i=1}^{n} f(a+i\Delta x)\,\Delta x$$

   (b) By interpreting the integral as an area
   (c) By the fundamental theorem of calculus (using an antiderivative)
   (d) By a substitution which changes the integral to one we can integrate
   (e) By special methods of integration, such as integration by parts, partial fractions, and trigonometric substitutions
   (f) By use of limits to evaluate improper integrals
   (g) Find approximations using the trapezoid rule
   (h) Find approximations using Maclaurin series
6. How to Use Integrals
   (a) To find areas of regions by horizontal or vertical strips
   (b) To find volumes of solids of revolutions by discs, washers, or shells
   (c) To find the volume of a general solid by use of cross-sections
   (d) To find the length of a curve
   (e) To find the surface area of a solid of revolution
   (f) To find the center of mass of a region
7. How to Find and Use Values for Convergent Series
   (a) Finding patterns for sequences, and using the sequence of partial sums concept to define an infinite series
   (b) Methods to test series for convergence or divergence
      i. Comparison test
      ii. Integral test
      iii. Limit form of the comparison test
      iv. Ratio test
      v. Root test
      vi. Alternating series test
      vii. Absolute convergence test
   (c) Methods to find the value for a convergent series (exact or approximate)
      i. Geometric series
      ii. Telescoping principle
      iii. Use of an improper integral
      iv. Use of terms from the sequence of partial sums
   (d) The use of Maclaurin series of functions to find approximations for various expressions

Section II. Extending Calculus Results for a Function of One Variable to Functions of More than One Variable

8. Helpful Tools to Study Such Functions
   (a) Parametric representation of curves
   (b) Polar coordinates as an alternative to rectangular coordinates
   (c) Vectors in 2 and 3 dimensions, along with the dot product and cross product
   (d) Equations of curves and surfaces. The straight line is the simplest curve and the plane is the simplest of the quadric surfaces, which are the 3-dimensional generalization of conics from 2 dimensions.
9. Extensions of the Derivative
   (a) The partial derivative—the basic tool for calculations. It is used to express all the other concepts below.
   (b) Directional derivative (partial derivatives are a special case)
   (c) Chain rule and implicit differentiation
   (d) Gradient vector—a generalization of the normal line to a curve
   (e) Tangent plane—a generalization of the tangent line to a curve
   (f) Max-min values for surfaces
10. Extensions of the Integral
   (a) Definitions of a double and triple integral
   (b) Iterated integrals—the basic tool for evaluation of double and triple integrals
   (c) Use of polar coordinates—a generalization of the substitution method to evaluate a single integral
   (d) New ways to represent volumes of solids
   (e) New applications such as mass of a solid
   (f) Line integral concept, which generalizes the definite integral $\int_a^b f(x)\,dx$
   (g) Major integral theorems, such as Green's theorem, Stokes's theorem, and the divergence theorem, each of which can be considered as a generalization of the fundamental theorem of calculus
11. How to Classify and Solve Differential Equations

   This topic comes after partial derivatives are introduced because the solution for an exact differential equation needs this concept. Otherwise, much of this topic could occur at the end of Section I.

   (a) Types of differential equations of the first order and first degree
      i. Separation of variables
      ii. Homogeneous
      iii. Integrating factors
      iv. Exact
      v. Linear

(b) Higher-order linear homogeneous differential equations with constant coefficients

(c) Higher-order linear non-homogeneous differential equations with constant coefficients

i. Method of undetermined coefficients

ii. Method of variation of parameters

(d) Higher-order differential equations with non-constant coefficients

i. Use a substitution to obtain a first-order, first-degree differential equation if possible

ii. Find a series solution by solving for the coefficients of $\sum_{n=1}^{\infty} a_n x^n$

Section III. Precalculus Material that Is Needed for Calculus

12. Basic Algebraic Techniques

(a) Factor polynomials and find zeroes

(b) Completing the square

(c) Multiply by conjugate factor, i.e.,

$$\frac{1}{\sqrt{a}-\sqrt{b}} = \frac{\sqrt{a}+\sqrt{b}}{a-b}$$

(d) Use of the binomial expansion $(a+b)^n$

(e) A partial fraction decomposition, i.e.,

$$\frac{2x+1}{x(x-1)^2} = \frac{A}{x} + \frac{B}{x-1} + \frac{C}{(x-1)^2}$$

13. Equations and Graphs of Familiar Curves

(a) The straight line in two dimensions

(b) The conics (parabola, circle, ellipse, and hyperbola)

(c) Radical functions (i.e., $\sqrt{x}$, $\sqrt[3]{x}$)

(d) Rational functions, i.e.,

$$\frac{2x+1}{x-2}$$

(e) The six trigonometric functions

(f) The inverse trigonometric functions, especially $\arcsin x$ and $\arctan x$

(g) The exponential and logarithmic functions ($y = b^x$ and $y = \log_b x$)

14. Basic Trigonometric Identities

(a) $\sin^2 x + \cos^2 x = 1$

(b) $\tan^2 x + 1 = \sec^2 x$

(c) $\sin(x+y) = \sin x \cos y + \cos x \sin y$

(d) $\cos(x + y) = \cos x \cos y - \sin x \sin y$

(e) $\sin 2x = 2 \sin x \cos x$

(f) $\cos 2x = \cos^2 x - \sin^2 x = 2\cos^2 x - 1 = 1 - 2\sin^2 x$

(g) $\sin^2 x = \dfrac{1 - \cos 2x}{2}$; $\cos^2 x = \dfrac{1 + \cos 2x}{2}$

(h) $\sin \dfrac{x}{2} = \sqrt{\dfrac{1 - \cos x}{2}}$; $\cos \dfrac{x}{2} = \sqrt{\dfrac{1 + \cos x}{2}}$

15. Basic Logarithmic and Exponential Identities

(a) $b^{x+y} = b^x b^y$

(b) $(b^x)^r = b^{rx}$

(c) $b^{x-y} = \dfrac{b^x}{b^y}$

(d) $\log_b (xy) = \log_b x + \log_b y$

(e) $\log_b x^r = r \log_b x$

(f) $\log_b \left(\dfrac{x}{y}\right) = \log_b x - \log_b y$

(g) $\log_b x = \dfrac{\log_a x}{\log_a b}$

(h) $\log_b b^x = x$

(i) $b^{\log_b x} = x$

16. Oft-used Formulas that are NOT True

(a) $\sqrt{a + b} = \sqrt{a} + \sqrt{b}$

(b) $\dfrac{1}{a + b} = \dfrac{1}{a} + \dfrac{1}{b}$

(c) $\sin 2x = 2 \sin x$

(d) $\log(x + y) = \log x + \log y$

(e) $\arcsin x = \dfrac{1}{\sin x}$

Calculus Review Problems

Introduction

The best preparation for the study of calculus is a strong background in pre-calculus material, including such topics as algebraic techniques, trigonometric identities, Cartesian geometry, and properties of functions. This is because calculus adds new ideas—such as limits, derivatives, and integrals—to the material of pre-calculus mathematics, which allow us to obtain new insights. Solutions of most calculus problems contain a significant amount of pre-calculus material.

In a similar way, success in the study of real analysis depends strongly on a good understanding of calculus. Real analysis is deeply involved with theoretical issues, yet it deals with the topics of limits, continuity, derivatives, integrals, and series that make up the subject matter of calculus. Most students begin their study of real analysis one year or more after completing their study of calculus and find their knowledge of and facility with calculus significantly diminished. For this reason, a review of calculus is excellent preparation for the real analysis course. The material below is intended to help you review the main topics from calculus that are needed for real analysis. Whereas they are divided into sections, some problems require results from more than one section. For example, the problems in Section 2 which use L'Hospital's rule need derivative formulas from Section 3.

Most of the problems include comments to describe the importance of the problem or provide some hint for its solution. You may wish to attempt the problem first and look at the comments after your initial attempt.

Section 1. Pre-calculus Material

1. (a) Give a reason to explain why

$$\sum_{i=1}^{n} i = \frac{n(n+1)}{2}.$$

(b) Prove the result in part (a) by the method of mathematical induction.

Comments: We use the summation formulas $\sum_{i=1}^{n} i^k$ at the beginning of the chapter on integration in calculus to illustrate the integral definition for simple functions of the form $f(x) = x^k$. In Example 5 of Problem 2 in Part II, the usual method for obtaining these formulas is described for the case when $k = 2$. It is a recursive approach, requiring knowledge of the formulas for $\sum_{i=1}^{n} i^t$ for all natural numbers $t < k$ in order to find the formula for $\sum_{i=1}^{n} i^k$. It is clear that

$$\sum_{i=1}^{n} i^0 = 1 + 1 + \cdots + 1 = n.$$

The formula

$$\sum_{i=1}^{n} i = 1 + 2 + \cdots + n = \frac{n(n+1)}{2}$$

is usually proved by writing

$$S = 1 + 2 + \cdots + (n-1) + n$$

and also in the reverse order as

$$S = n + (n-1) + \cdots + 2 + 1.$$

We obtain the desired formula by adding the two expressions together and simplifying.

2. Show that the slope of the line between the two points on the graph of $f(x) = \sqrt{x}$ determined by $x = a$ and $x = b$ (where $0 < a < b$) is equal to

$$\frac{1}{\sqrt{a} + \sqrt{b}}.$$

This line is sometimes referred to as the secant line determined by $x = a$ and $x = b$.

3. Nine familiar trigonometric identities are listed below. Assume that the first three are true, and use them to prove the next six.

$$\sin^2 x + \cos^2 x = 1 \tag{1}$$

$$\sin(x + y) = \sin x \cos y + \cos x \sin y \tag{2}$$

$$\cos(x + y) = \cos x \cos y - \sin x \sin y \tag{3}$$

$$\cos 2x = \cos^2 x - \sin^2 x \tag{4}$$

$$\sin^2 x = \frac{1 - \cos 2x}{2} \tag{5}$$

$$\cos^2 x = \frac{1 + \cos 2x}{2} \tag{6}$$

$$\sin\frac{x}{2} = \sqrt{\frac{1 - \cos x}{2}} \tag{7}$$

$$\tan^2 x + 1 = \sec^2 x \tag{8}$$

$$\tan(x + y) = \frac{\tan x + \tan y}{1 - \tan x \tan y} \tag{9}$$

Comments: Trigonometric identities occur regularly in calculus. For example, identities (1) and (8) are used to simplify many calculations, identity (2) is used for the proof of the derivative formula for $\sin x$, and identities (5) and (6) are used to evaluate integrals such as

$$\int \sin^2 x\,dx \quad \text{and} \quad \int \cos^2 x\,dx.$$

4. Let t represent an angle using radian measure in a circle with radius 1 that is centered at the origin, as shown in Figure 1. It is clear from the graph that

$$\text{Area of } \triangle COA < \text{Area of sector } COB < \text{Area of } \triangle DOB.$$

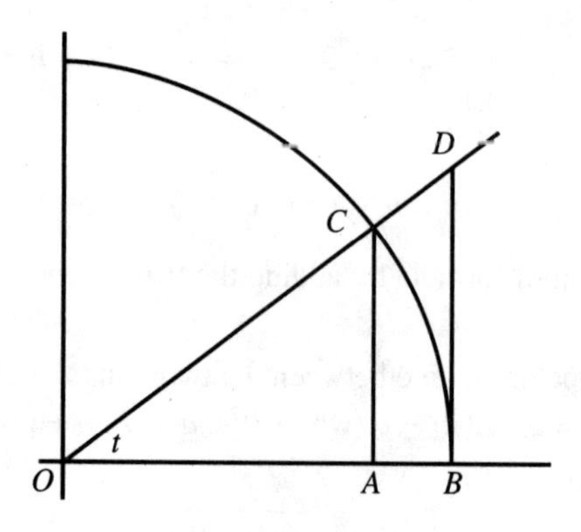

FIGURE 1

Substitute an expression for each of these three areas into this inequality and verify that it can be rewritten as

$$\cos t < \frac{\sin t}{t} < \frac{1}{\cos t}.$$

Then explain how the squeeze theorem is used to prove that

$$\lim_{t \to 0^+} \frac{\sin t}{t} = 1.$$

Comments: The result that

$$\lim_{t \to 0} \frac{\sin t}{t} = 1$$

is very important for the derivation of such calculus formulas as

$$\frac{d}{dx} \sin x = \cos x.$$

All that is needed to prove

$$\lim_{t \to 0} \frac{\sin t}{t} = 1$$

are some simple trigonometric identities and the squeeze theorem.

Section 2. Limits and Continuity

5. (a) Verify that the function $f(x) = x \sin(1/x)$ if $x \neq 0$ has a removable discontinuity at $x = 0$.
   (b) Verify that the continuous function in part (a) (assuming that $f(0) = 0$) is not differentiable at $x = 0$.

   Comments: The result in part (a) can be verified by use of the squeeze theorem. The result in part (b) can be shown by use of the definition of $f'(0)$.

6. Evaluate

$$\lim_{h \to 0} \frac{\sqrt{2+h} - \sqrt{2}}{h}$$

   by the following three methods.
   (a) Use a conjugate factor.
   (b) Use L'Hospital's rule.
   (c) Use the derivative of $f(x) = \sqrt{2+x}$, along with the definition of $f'(0)$.

7. (a) Use L'Hospital's rule to evaluate

$$\lim_{x \to \infty} \left(1 + \frac{a}{x}\right)^{bx}.$$

(b) How is the problem in part (a) related to the problem of finding the limit of the sequence $\{x_n\}$ where

$$x_n = \left(1 + \frac{a}{n}\right)^{bn}?$$

Comments: The limit in part (a) is indeterminate of the form 1^∞, so we first set

$$y = \left(1 + \frac{a}{x}\right)^{bx}$$

and then rewrite this in the form

$$\ln y = \frac{b\ln(1 + \frac{a}{x})}{\frac{1}{x}}$$

before applying L'Hospital's rule. The function

$$f(x) = \left(1 + \frac{a}{x}\right)^{bx}$$

in part (a) is sometimes referred to as the "related function" for the sequence in part (b), which has the same formula as the function in part (a), but is only defined for natural numbers n. The same idea is used in the integral test for positive series. For example, to test the series

$$\sum_{n=1}^{\infty} \frac{1}{n^2},$$

consider the improper integral

$$\int_1^{\infty} \frac{1}{x^2}\, dx.$$

In this case,

$$f(x) = \frac{1}{x^2}$$

is the related function for the sequence $\{a_n\}$ where

$$a_n = \frac{1}{n^2}.$$

Section 3. Derivatives

8. Show how to use the formula

$$\frac{d}{dx}(\cos x) = -\sin x$$

in order to obtain the usual derivative formulas for $\sec x$ and $\arccos x$.

Comments: This problem illustrates that there is a natural order in obtaining the derivative formulas. We can use the derivative formula for $\cos x$ along with the quotient rule to obtain the derivative formula for $\sec x$. We can use the derivative formula for $\cos x$ along with the chain rule to obtain the derivative formula for $\arccos x$, noting also that $y = \arccos x$ is equivalent to $x = \cos y$.

9. (a) Show how to use the definition of the derivative in order to obtain the power rule formula that

$$\frac{d}{dx} x^r = r x^{r-1}$$

when r is a positive integer.

(b) Show how to establish this formula if r is a negative integer.

(c) Show how to establish this formula if r is a rational number.

(d) Show how to establish this formula if r is an irrational number.

Comments: While the power rule formula is true for all real numbers r, the proof of this result occurs in several stages during a calculus course. The first result in part (a) is obtained by using the definition of the derivative and the binomial expansion $(a+b)^n$. The result in part (b) is found by using the substitution $r = -n$, along with the quotient rule and the result in part (a). The result in part (c) is obtained using the substitution $r = p/q$, where p and q are integers, the observation that $y = x^{p/q}$ is equivalent to $y^q = x^p$, the chain rule, and the result in part (b). The general result in part (d) is proved after the natural logarithm function and its derivative is introduced. We observe here that $y = x^r$ is equivalent to $\ln y = r \ln x$, and then differentiate both sides.

10. Some special techniques for finding derivatives include the method of implicit differentiation, the method of logarithmic differentiation, the use of an inverse function, and the use of a function defined by an integral. Use each of these methods once to find dy/dx in the following cases.

(a) $xy^3 + e^y = 2x$

(b) $y = \left(1 + \frac{a}{x}\right)^{bx}$

(c) $y = f^{-1}(x)$, where $f(x) = e^{\arctan x} - 2$

(d) $y = \int_{x^2}^{2} \sqrt[3]{1 - t^2}\, dt$

11. (a) When Newton's method, i.e., that

$$x_{n+1} = x_n - \frac{f'(x_n)}{f(x_n)}$$

is used to approximate $\sqrt{a}$, show that the formula for x_{n+1} can be written in the form

$$\frac{1}{2}\left(x_n + \frac{a}{x_n}\right).$$

This determines a recursively defined sequence, a topic of importance early in real analysis.

(b) Find the first four terms for the sequence in part (a) when $a = 3$ and $x_1 = 1$.

Comments: To approximate $\sqrt{a}$ by Newton's method, which is used to find the zero of a function, consider the function $f(x) = x^2 - a$. In part (b), you should find that $x_4 = \frac{97}{56} \approx 1.7321$.

12. Find the value for c so that the tangent line to $f(x) = \sqrt{x}$ at the point $(c, f(c))$ is parallel to the secant line determined by $x = a$ and $x = b$. Assume that $0 < a < b$. Recall that the secant line is the straight line between the points $(a, f(a))$ and $(b, f(b))$.

Comments: One way to solve this problem is to apply the mean value theorem to $f(x) = \sqrt{x}$ on the interval $[a, b]$. The mean value theorem is an important result in real analysis, and is used in the proof of the fundamental theorem of calculus.

Section 4. Integration

13. Without integrating, show how to verify that

$$\int \tan^4 x\, dx = \frac{1}{3}\tan^3 x - \tan x + x.$$

Comments: The usual method to integrate $\int \tan^4 x\, dx$ is by use of the identity that $\tan^4 x = \tan^2 x(\sec^2 x - 1)$. But in this case, you are to use the fact that $\int f(x)\, dx = F(x)$ if $F'(x) = f(x)$.

14. (a) Show that the function

$$f(x) = \frac{\ln x}{x}$$

has a relative maximum at the point $(e, 1/e)$.

(b) Find the interval over which the graph of $f(x)$ is concave up.

(c) Sketch a graph of $f(x)$, illustrating the properties found in parts (a) and (b).

(d) Show the area of the region under the graph of $f(x)$ and above the interval $[1, e]$ is $\frac{1}{2}$.

15. (a) Find an expression for the sequence term S_n which is used to express the Riemann sum

$$S_n = \sum_{i=1}^{n} f(c_i)\Delta x$$

for approximating the value of the definite integral

$$\int_2^5 x^3\, dx.$$

Assume the interval [2, 5] is divided into n equal parts and the right endpoint is used for the c_i value. Sketch a graph showing the n rectangles involved in this approximation.

(b) Find the limit of the sequence in part (a) and check that it gives the expected value for

$$\int_2^5 x^3\,dx.$$

Hint:

$$\sum_{i=1}^{n} i^3 = \frac{n^2(n+1)^2}{4}.$$

Comments: In part (a), you should use

$$\Delta x = \frac{3}{n} \quad \text{and} \quad c_i = 2 + \frac{3i}{n}.$$

16. (a) Use L'Hospital's rule to show that $f(x) = x \ln |x|$ has a removable discontinuity at $x = 0$ if we define $f(0) = 0$.

(b) Use integration by parts to show that

$$\int x \ln x\,dx = \frac{x^2 \ln x}{2} - \frac{x^2}{4}.$$

Then explain why

$$\int_0^1 x \ln x\,dx = -\frac{1}{4}.$$

Comments: In Problem 17 in Part II, we extend the above result to show

$$\lim_{x\to 0+} x^m (\ln x)^n = 0 \quad \text{for all } m > 0, \text{ and } n > 0$$

and also that

$$\int_0^1 (x \ln x)^n\,dx = \frac{(-1)^{n-1}\,n!}{(n+1)^{n+1}}.$$

17. (a) Use integration by parts twice to show that

$$\int_{-\pi}^{\pi} x^2 \cos nx\,dx = \frac{(-1)^n 4\pi}{n^2}.$$

(b) Use integration by parts to establish the following two recursion formulas. Then use them to evaluate the integral in part (a).

$$\int x^k \cos nx\, dx = \frac{x^k \sin nx}{n} - \frac{k}{n}\int x^{k-1} \sin nx\, dx$$

$$\int x^k \sin nx\, dx = -\frac{x^k \cos nx}{n} + \frac{k}{n}\int x^{k-1} \cos nx\, dx$$

Comments: The integral in part (a) is needed to find the Fourier series for $f(x) = x^2$. The recursion formulas are especially useful when several integrations by parts are needed, such as for the integral $\int x^4 \sin nx\, dx$.

18. Find the value for $\int_0^1 x^p\, dx$ for all real values of p. You will need to consider different cases depending on the value of p.

Comments: For $p \geq 0$, this is a proper definite integral, while for $p < 0$, it is an improper integral.

19. Use the trapezoid rule with $n = 4$ to approximate the value of the integral $\int_0^1 x^x\, dx$.

Comments: The fundamental theorem does not apply to this integral because there is no known antiderivative. The trapezoid rule gives one method for approximating such an integral. An alternate method using a Maclaurin series is described in Problem 18 in Part II.

20. Change the double integral below to an equivalent double integral in polar coordinates and then evaluate it to obtain the answer of $\pi/4$.

$$\int\!\!\int_R e^{-(x^2+y^2)}\, dx\, dy \text{ where the region } R \text{ is the first quadrant.}$$

Comments: All previous problems have been limited to functions of one variable because our goal is to review material needed for the study of functions of one real variable. However this double integral is needed in Problem 29 in Part II to find a value for the gamma function $\Gamma(x)$ when $x = \frac{1}{2}$.

Section 5. Infinite Series

21. (a) Show the value is $\frac{3}{2}$ for the geometric series that begins $2 - \frac{2}{3} + \cdots$.
    (b) Find the next two terms in this series.
22. Use the technique of partial fraction decomposition to evaluate the following quantities:
    (a) the integral

$$\int_1^2 \frac{1}{x(x+2)}\, dx$$

    (b) the series

$$\sum_{n=1}^{\infty} \frac{1}{n(n+2)}$$

Comments: Note the similarity in the method of solution for these two problems. For the integral in part (a), we use the partial fraction decomposition determined by the identity

$$\frac{1}{x(x+2)} = \frac{1}{2}\left(\frac{1}{x} - \frac{1}{x+2}\right)$$

to show that the value of the integral is $\frac{1}{2}\ln(1.5)$. For the series in part (b), we use the identity

$$\frac{1}{n(n+2)} = \frac{1}{2}\left(\frac{1}{n} - \frac{1}{n+2}\right)$$

and the cancellation principle to show that the series converges to $\frac{3}{4}$.

23. (a) Evaluate the improper integral

$$\int_3^\infty \frac{1}{x^2}\,dx.$$

(b) Use the result in part (a) and the inequalities given below to find upper and lower bounds for the series

$$\sum_{n=1}^{\infty} \frac{1}{n^2}.$$

Comments: Improper integrals of the form $\int_k^\infty f(x)\,dx$ are used not only to verify convergence or divergence of the series

$$\sum_{n=1}^{\infty} f(n)$$

by use of the integral test, but also to provide upper and lower bounds for the value of a convergent series. If the sequence of partial sums is denoted by

$$S_n = \sum_{i=1}^{n} f(i),$$

then the general result is

$$S_{k-1} + \int_k^\infty f(x)\,dx < \sum_{n=1}^{\infty} f(n) < S_k + \int_k^\infty f(x)\,dx.$$

In this problem, you should find that

$$\frac{19}{12} < \sum_{n=1}^{\infty} \frac{1}{n^2} < \frac{61}{36}.$$

24. Test the series

$$\sum_{n=1}^{\infty} \frac{\ln n}{n^2}$$

by the following four methods.

(a) Show the ratio test fails when applied to this series.
(b) Show the integral test implies convergence.
(c) Explain why the limit comparison test fails when using both

$$\sum \frac{1}{n} \quad \text{and} \quad \sum \frac{1}{n^2}$$

as the comparison series.

(d) Explain why the limit comparison test implies convergence with

$$\sum \frac{1}{n^{3/2}}$$

as the comparison series.

Comments: This problem reviews the standard tests for convergence of a series of positive terms. In order to show the ratio test fails to apply, we need to show

$$\lim_{n\to\infty} \frac{a_{n+1}}{a_n} = 1,$$

where

$$a_n = \frac{\ln n}{n^2}.$$

In order to show the series converges by the integral test, we need to show

$$\int_1^{\infty} \frac{\ln x}{x^2}\, dx$$

is a convergent improper integral, which can be done by the use of integration by parts. The limit comparison test will fail with the test series of

$$\sum_{n=1}^{\infty} \frac{1}{n} \quad \text{if} \quad \lim_{n\to\infty} \frac{a_n}{\frac{1}{n}} = 0.$$

The limit comparison test will imply convergence if the test series

$$\sum_{n=1}^{\infty} b_n$$

converges and

$$\lim_{n\to\infty} \frac{a_n}{b_n} \neq \infty.$$

25. (a) Show how to verify that the first four terms for the Maclaurin series of $f(x) = \sqrt{x+1}$ are $1 + \frac{1}{2}x - \frac{1}{8}x^2 + \frac{1}{16}x^3$. This is also called the Taylor polynomial of degree 3 for $f(x) = \sqrt{x+1}$ with $a = 0$.

    (b) Use your result in part (a) to find an approximation for $\sqrt{3}$. Compare it with the value found earlier by Newton's method in Problem 11.

Comments: The usual method to find the Maclaurin series for $f(x)$ is to calculate the values for $f^{(n)}(0)$ and substitute them into the formula

$$f(x) = \sum_{n=0}^{\infty} \frac{f^{(n)}(0)}{n!} x^n.$$

For part (b), we cannot get an approximation for $\sqrt{3}$ by substituting $x = 2$ into the series in part (a) because the interval of convergence for that series is only $-1 < x < 1$. A valid choice is $x = -\frac{2}{3}$ because

$$\sqrt{-\tfrac{2}{3}+1} = \sqrt{\tfrac{1}{3}} = \tfrac{1}{3}\sqrt{3},$$

and the Taylor polynomial of degree 3 for this value gives $\sqrt{3} \approx 1.7$. A better choice is $x = \frac{1}{3}$ because

$$\sqrt{\tfrac{1}{3}+1} = \tfrac{2}{3}\sqrt{3}$$

and the Taylor polynomial of degree 3 for this value gives $\sqrt{3} \approx 1.7326$.

II

Analysis Problems

Introduction

The following collection of 34 problems is divided into three sections. Ample hints and examples are provided to guide you through each problem. Where appropriate, historical background is provided, or comments are included to describe the setting and importance of the problem.

Section 1 consists of four basic problems from calculus that may profitably be studied prior to an analysis course. They demonstrate some of the underlying patterns among the fundamental topics of calculus, such as the derivative, the integral, and sequences and series.

The seventeen problems in Section 2 supplement material presented in a traditional course in real analysis. Some of the problems such as #8, 9 and 20 are in the form of a chart to be filled in. Other problems, such as #10, 11 and 12, present a variety of ways to solve the same problem.

Section 3 contains thirteen problems which can be used for enrichment purposes. These enable the student to gain an introduction to a topic that might not otherwise be covered, such as the gamma function, Bernoulli numbers, Fourier series, and Lebesgue measure. I have assigned these to be done as a project by a team of students who write up their results and make a presentation to the class.

Section 1. Basic Problems

Problem 1. Properties of Even and Odd Functions

This problem about properties of even and odd functions is helpful in two ways. First, it gives practice in working with some basic definitions from calculus, and second, it provides shortcuts for the evaluation of definite integrals that occur in settings such as the determination of the Fourier series for a function. We begin with some definitions.

Definition 1. f is an *even function* if $f(-x) = f(x)$ for x in the domain of f.

Definition 2. f is an *odd function* if $f(-x) = -f(x)$ for x in the domain of f.

Definition 3. Two functions f and g have *opposite parity* if one is even and the other is odd. f and g have the *same parity* if both are even or both are odd.

We only use the term parity for functions that are either even or odd. The functions x^3 and $\sin x$ are odd functions, while x^4 and $\cos x$ are even functions. Some functions as e^x or $\sqrt{x}$ or $x + x^2$ are neither even nor odd. The following results assume the functions under discussion are either even or odd.

Theorem 1. *The product $f \cdot g$ is odd if and only if f and g have opposite parity.*

Proof The proof is straightforward, involving several cases such as if f is odd and g is even, then

$$(f \cdot g)(-x) = f(-x)g(-x) = -f(x)g(x) = -(f \cdot g)(x),$$

so $f \cdot g$ is odd. The remaining cases are left as an exercise. ■

Corollary 2. *The quotient f/g is odd if and only if f and g have opposite parity.*

Proof Observe first that g and its reciprocal $1/g$ have the same parity. If f and g are both odd, then $1/g$ is odd, and $f/g = f \cdot (1/g)$ is even by Theorem 1. The rest of this proof is left as an exercise. ■

Example 1. The function $\sec x$ is even, being the reciprocal of the even function $\cos x$, while $\tan x$ is odd, being the quotient of the odd function $\sin x$ and the even function $\cos x$. The Maclaurin series of an odd function has only odd powers of x, while the Maclaurin series of an even function has only even powers of x.

Theorem 3. *If f is differentiable, then f and f' have opposite parity.*

Proof Suppose f is an odd function and is differentiable at $x = \pm a$. Then f' is an even function because

$$\begin{aligned} f'(a) &= \lim_{x \to a} \frac{f(x) - f(a)}{x - a} \\ &= \lim_{-u \to a} \frac{f(-u) - f(a)}{-u - a} \\ &= \lim_{u \to -a} \frac{-f(u) + f(-a)}{-(u + a)} \\ &= \lim_{u \to -a} \frac{f(u) - f(-a)}{u - (-a)} \\ &= f'(-a). \end{aligned}$$

The case where f is an even function is left for an exercise. ■

Corollary 4. *If f has an antiderivative F, then f and F have opposite parity.*

Proof This is left as an exercise. ■

Theorem 5. *If f is integrable on $[-a, a]$, then*

$$\int_{-a}^{a} f(x)\,dx = \begin{cases} 2\int_0^a f(x)\,dx & \text{if } f \text{ is even} \\ 0 & \text{if } f \text{ is odd.} \end{cases}$$

Proof If f is even, then

$$\begin{aligned} \int_{-a}^{a} f(x)\,dx &= \int_{-a}^{0} f(x)\,dx + \int_0^a f(x)\,dx \\ &= \int_a^0 f(-u)(-du) + \int_0^a f(u)\,du \quad \text{by substitution} \\ &= \int_0^a f(u)\,du + \int_0^a f(u)\,du \\ &= 2\int_0^a f(x)\,dx. \end{aligned}$$

The case when f is odd is left for the exercises. ■

Example 2. The following integral formulas are helpful when working with Fourier series (see Problem 31). The first three formulas are a direct application of Theorem 5, while formulas (4)–(7) need trig identities for their derivation (see Exercise 6). We assume n and k are positive integers.

$$\int_{-\pi}^{\pi} x^k \sin nx\,dx = 0 \quad \text{if } k \text{ is even} \tag{1}$$

$$\int_{-\pi}^{\pi} x^k \cos nx\,dx = 0 \quad \text{if } k \text{ is odd} \tag{2}$$

$$\int_{-\pi}^{\pi} \sin nx \cos kx\,dx = 0 \quad \text{for all } n, k \tag{3}$$

$$\int_{-\pi}^{\pi} \sin nx \sin kx\,dx = 0 \quad \text{for all } n \neq k \tag{4}$$

$$\int_{-\pi}^{\pi} \sin^2 nx\,dx = \pi \tag{5}$$

$$\int_{-\pi}^{\pi} \cos nx \cos kx\,dx = 0 \quad \text{for } n \neq k \tag{6}$$

$$\int_{-\pi}^{\pi} \cos^2 nx\,dx = \pi \tag{7}$$

Theorem 6. *If f is odd and f^{-1} exists, then f^{-1} is also odd.*

Proof Hint: Let $u = -f^{-1}(-x)$ and show that $u = f^{-1}(x)$. ■

Exercises

1. Finish the proof of Theorem 1 by deciding which additional cases are needed, and then proving each one.
2. Complete the proof of Corollary 2.
3. Prove Theorem 3 with the assumption that f is an even function.
4. Prove Corollary 4.
5. Adjust the proof of Theorem 5 for the case when f is odd.
6. Use the following trig identities to prove integration formulas (4)–(7) above.

$$\sin nx \sin kx = \tfrac{1}{2}\left(\cos(n-k)x - \cos(n+k)x\right)$$
$$\cos nx \cos kx = \tfrac{1}{2}\left(\cos(n-k)x + \cos(n+k)x\right)$$

7. Use the hint given to complete a proof of Theorem 6.
8. Explain why there is no extension of the result in Theorem 6 for even functions.

Problem 2. Use of the Cancellation Principle

A helpful way to get the "big picture" in analysis is to identify some common ideas or techniques that are used repeatedly to obtain a variety of results. One of these is the *cancellation principle* where most of the terms cancel out to reduce a complicated expression to a much simpler form. We next present several examples of this. Fill in the details for each one and add additional examples if you can.

Example 1. The formula for a geometric series $\sum_{n=0}^{\infty} r^n$.

Let $S_n = 1 + r + r^2 + \cdots + r^{n-1}$. Then look at $S_n - r\,S_n$ and use the *cancellation principle* to show that

$$S_n = \frac{1-r^n}{1-r},$$

from which we get

$$\lim_{n\to\infty} S_n = \frac{1}{1-r} \quad \text{if} \quad |r| < 1.$$

Example 2. The telescoping principle for infinite series.

Use the identity

$$\frac{1}{n(n+1)} = \frac{1}{n} - \frac{1}{n+1}$$

and the *cancellation principle* to show

$$\sum_{n=1}^{\infty} \frac{1}{n(n+1)} = 1.$$

This idea can be extended to series of the form

$$\sum_{n=1}^{\infty} \frac{1}{n(n+k)} \quad \text{or} \quad \sum_{n=1}^{\infty} \frac{1}{n(n+j)(n+k)}.$$

Example 3. If

$$S = \sum_{n=1}^{\infty} \frac{1}{n^2} = \frac{\pi^2}{6}$$

is assumed, and

$$R = \sum_{n=1}^{\infty} \frac{(-1)^{n+1}}{n^2},$$

find $S - R$ and use the *cancellation principle* to show that

$$R = \frac{1}{2}S = \frac{\pi^2}{12}.$$

Example 4. In the proof of the fundamental theorem of calculus, the goal is to find the value of $\int_a^b f(x)\,dx$, given that $F'(x) = f(x)$. We begin with $P = \{x_i\}_{i=0}^n$, which is an arbitrary partition of the interval $[a, b]$. By applying the mean value theorem to $F(x)$ on each subinterval $[x_{i-1}, x_i]$, we get

$$F(x_i) - F(x_{i-1}) = f(c_i)\Delta x_i \quad \text{for } i = 1, 2, \ldots, n \quad \text{and for some } c_i \in (x_{i-1}, x_i).$$

Observe that the Riemann sum $\sum_{i=1}^n f(c_i)\Delta x_i$ is equal to

$$\sum_{i=1}^{n} (F(x_i) - F(x_{i-1})),$$

which can be reduced to $F(b) - F(a)$ by use of the *cancellation principle*. It then follows that $\int_a^b f(x)\,dx = F(b) - F(a)$.

Example 5. In order to obtain the summation formulas for $\sum_{i=1}^n i^k$, the expression

$$(i+1)^{k+1} - i^{k+1}$$

is used along with the *cancellation principle*. For example, to find the formula for $\sum_{i=1}^{n} i^2$, we first establish

$$\sum_{i=1}^{n} 1 = n \quad \text{and} \quad \sum_{i=1}^{n} i = \frac{n(n+1)}{2},$$

and then observe that

$$(i+1)^3 - i^3 = 3i^2 + 3i + 1.$$

By substituting $i = 1, i = 2, \ldots, i = n$ into this formula, we obtain the following.

$$\begin{aligned}
2^3 - 1^3 &= 3 \cdot 1^2 + 3 \cdot 1 + 1 \\
3^3 - 2^3 &= 3 \cdot 2^2 + 3 \cdot 2 + 1 \\
4^3 - 3^3 &= 3 \cdot 3^2 + 3 \cdot 3 + 1 \\
&\vdots \\
(n+1)^3 - n^3 &= 3 \cdot n^2 + 3 \cdot n + 1
\end{aligned}$$

By adding the columns on the right side and using the *cancellation principle* on the left side, we obtain

$$(n+1)^3 - 1^3 = 3 \sum_{i=1}^{n} i^2 + 3 \sum_{i=1}^{n} i + \sum_{i=1}^{n} 1.$$

After substituting and simplifying, we finally get the desired formula

$$\sum_{i=1}^{n} i^2 = \frac{n(n+1)(2n+1)}{6}.$$

These summation formulas are encountered at the beginning of the integration chapter in calculus and are discussed further in Problem 4.

Exercises

1. Explain why the geometric series $\sum_{n=0}^{\infty} r^n$ diverges if $|r| \geq 1$.

The next three problems generalize the result in Example 2.

2. Use the cancellation principle to show that

$$\sum_{n=1}^{\infty} \frac{1}{n(n+2)} = \frac{3}{4}.$$

Begin with the identity

$$\frac{1}{n(n+2)} = \frac{1}{2}\left(\frac{1}{n} - \frac{1}{n+2}\right).$$

3. Establish the general formula

$$\sum_{n=1}^{\infty} \frac{1}{n(n+k)} = \frac{1}{k} \sum_{i=1}^{k} \frac{1}{i}.$$

4. Establish the formula

$$\sum_{n=1}^{\infty} \frac{1}{n(n+1)(n+3)} = \frac{7}{36}.$$

Hint: Begin by obtaining the partial fraction decomposition

$$\frac{1}{n(n+1)(n+3)} = \frac{1}{3} \cdot \frac{1}{n} - \frac{1}{2} \cdot \frac{1}{n+1} + \frac{1}{6} \cdot \frac{1}{n+3}$$

and rewrite the right-hand side as

$$\frac{1}{3}\left(\frac{1}{n} - \frac{1}{n+1}\right) - \frac{1}{6}\left(\frac{1}{n+1} - \frac{1}{n+3}\right).$$

The next two problems generalize the result in Example 3.

5. If

$$R = \sum_{n=1}^{\infty} \frac{(-1)^{n-1}}{n^3}, \quad S = \sum_{n=1}^{\infty} \frac{1}{n^3}, \quad \text{and} \quad T = \sum_{n=1}^{\infty} \frac{1}{(2n-1)^3},$$

show that $R = \frac{3}{4}S$ and $T = \frac{7}{8}S$.

6. If

$$R = \sum_{n=1}^{\infty} \frac{(-1)^{n-1}}{n^k}, \quad S = \sum_{n=1}^{\infty} \frac{1}{n^k}, \quad \text{and} \quad T = \sum_{n=1}^{\infty} \frac{1}{(2n-1)^k},$$

show that

$$R = \frac{2^{k-1}-1}{2^{k-1}} S \quad \text{and} \quad T = \frac{2^k - 1}{2^k} S.$$

7. Give reasons for each of the steps described in Example 4.

8. Complete the work outlined in Example 5 for the proof that

$$\sum_{i=1}^{n} i^2 = \frac{n(n+1)(2n+1)}{6}.$$

9. Assuming all the results given in Example 5 above, use the method described to find that

$$\sum_{i=1}^{n} i^3 = \frac{n^2(n+1)^2}{4}.$$

Begin with the expression $(i+1)^4 - i^4$. Refer to Problem 4 for further information on this general problem. After experiencing how much detail is needed to establish this formula for $\sum_{i=1}^{n} i^3$, one gains an appreciation (or a sense of pity) for the efforts of Cavalieri who established similar formulas in the 1630s for

$$\sum_{i=1}^{n} i^4, \sum_{i=1}^{n} i^5, \ldots, \sum_{i=1}^{n} i^9.$$

Problem 3. Finding Maclaurin Series Representations

In calculus, we show that if the function $f(x)$ can be written in the form

$$f(x) = a_0 + a_1 x + a_2 x^2 + a_3 x^3 + \cdots$$

and if all the derivatives of $f(x)$ are defined at $x = 0$, then the coefficients a_n must equal

$$\frac{f^{(n)}(0)}{n!}.$$

This expression

$$\sum_{n=0}^{\infty} \frac{f^{(n)}(0)}{n!} x^n$$

is called the Maclaurin series representation for $f(x)$. We present four methods for finding these series.

Method 1. The standard method of finding the Maclaurin series for $f(x)$ is to calculate the first several derivatives of $f(x)$, look for a pattern to express the general term $f^{(n)}(0)$, and then substitute these values into

$$f(x) = \sum_{n=0}^{\infty} \frac{f^{(n)}(0)}{n!} x^n.$$

This method works very well for functions such as $\sin x$, $\cos x$ and e^x because the derivatives have a recurring pattern. With additional effort, this method works also for the functions $\ln(x+1)$, $\sqrt{x+1}$ and $(x+1)^r$ for general values of r. However the standard method does not work well for functions such as $\tan x$, $\sec x$, $\arcsin x$, $\arctan x$, $\sin^2 x$ and $x/(e^x - 1)$ because no simple expression can be found for $f^{(n)}(0)$. Hence we look for alternate methods.

Method 2. Multiplication and Division of Known Series. One alternate method is to combine the formulas developed by the standard method with the algebraic processes of multiplication and division. For example, since

$$\tan x = \frac{\sin x}{\cos x} = \frac{x - \frac{x^3}{3!} + \frac{x^5}{5!} - + \cdots}{1 - \frac{x^2}{2!} + \frac{x^4}{4!} - + \cdots}$$

we can use long division to find the beginning terms of the Maclaurin series for $\tan x$. The same method works for $\sec x$ and $x/(e^x - 1)$. The series for $x/(e^x - 1)$ is important because it gives a definition for the Bernoulli numbers (see Problem 30). Notice that $\cot x$ and $\csc x$ have no Maclaurin series, since these functions have an infinite discontinuity at $x = 0$. The related technique of multiplication of known series can be used to find the series for $\sin^2 x$, as you are asked to do in Exercise 3.

Method 3. Substitution into a Known Series. We can replace x by $-x$ in the known series

$$e^x = 1 + x + \frac{x^2}{2!} + \frac{x^3}{3!} + \cdots$$

in order to obtain the new series

$$e^{-x} = 1 - x + \frac{x^2}{2!} - \frac{x^3}{3!} + - \cdots.$$

Or we can replace x by $2x$ in the known series

$$\sin x = x - \frac{x^3}{3!} + \frac{x^5}{5!} - + \cdots$$

in order to obtain the new series

$$\sin 2x = (2x) - \frac{1}{3!}(2x)^3 + \frac{1}{5!}(2x)^5 - + \cdots = 2x - \frac{4}{3}x^3 + \frac{4}{15}x^5 - + \cdots.$$

Leonhard Euler was bold enough to replace x in the series for e^x by the complex expressions ix and $-ix$, where $i = \sqrt{-1}$, in order to obtain such formulas as

$$e^{ix} = \cos x + i \sin x \quad \text{and} \quad \cos x = \frac{e^{ix} + e^{-ix}}{2}$$

(see Exercise 7).

Method 4. Differentiation and Integration of Known Series. Another useful method is to take the derivative or integral of a known series expansion in order to obtain a new Maclaurin series. For example, if $f(x) = \sin^2 x$, since

$$f'(x) = 2 \sin x \cos x = \sin 2x,$$

we can integrate term-by-term in the series for $\sin 2x$ as found by Method 3 to find the series for $\sin^2 x$ (see Exercise 5). One needs to also evaluate the integration constant (which is zero in this case) as the final step.

Similarly for $f(x) = \arctan x$, since

$$f'(x) = \frac{1}{1 + x^2},$$

we replace x by t^2 in the known geometric series

$$\frac{1}{1+x} = 1 - x + x^2 - x^3 + x^4 - + \cdots$$

and then integrate term-by-term to find the following series.

$$\begin{aligned}\arctan x &= \int_0^x \frac{1}{1+t^2}\,dt \\ &= \int_0^x (1 - t^2 + t^4 - t^6 + \cdots)\,dt \\ &= x - \frac{1}{3}x^3 + \frac{1}{5}x^5 - \frac{1}{7}x^7 + \cdots + \frac{(-1)^{n-1}x^{2n-1}}{2n-1} + \cdots.\end{aligned}$$

A similar technique applies to $f(x) = \arcsin x$ (see Exercise 6).

Exercises

1. Apply Method 1 to $f(x) = \cos x$ to show that

$$f^{(2n)}(0) = (-1)^n \quad \text{and} \quad f^{(2n-1)}(0) = 0.$$

2. **(a)** Apply Method 1 to $f(x) = \sqrt{x+1}$ to show that

$$f^{(n)}(0) = \frac{(-1)^{n-1} 1 \cdot 3 \cdots (2n-3)}{2^n}$$

when $n > 1$ and that

$$\begin{aligned}\sqrt{x+1} &= 1 + \frac{1}{2}x - \frac{1}{8}x^2 + \frac{1}{16}x^3 - \frac{5}{128}x^4 + \cdots + \\ &= 1 + \frac{1}{2}x + \sum_{n=2}^{\infty} \frac{(-1)^{n-1} 1 \cdot 3 \cdots (2n-3)}{2^n \cdot n!} x^n.\end{aligned}$$

(b) Use the ratio test to show that the power series in part (a) converges if $-1 < x < 1$, and that the ratio test fails to apply if $x = \pm 1$.

(c) When Isaac Newton found the above equation, he squared both sides as a check on his work. Calculate

$$\left(1 + \tfrac{1}{2}x - \tfrac{1}{8}x^2 + \tfrac{1}{16}x^3\right)^2$$

and discuss the meaning of your result.

3. Apply the technique of long division as described under Method 2 to show

$$\tan x = x + \tfrac{1}{3}x^3 + \tfrac{2}{15}x^5 + \cdots.$$

4. Apply the technique of multiplication of series to show that

$$\sin^2 x = \left(x - \frac{x^3}{3!} + \frac{x^5}{5!} - \frac{x^7}{7!} + - \cdots\right)^2 = x^2 - \frac{1}{3}x^4 + \frac{2}{45}x^6 - \frac{1}{315}x^8 + - \cdots.$$

5. Apply the technique described in Method 4 to obtain the same series for $\sin^2 x$ that was found in Exercise 4.

6. **(a)** Apply Method 4 to find the series expansion for

$$f(x) = \arcsin x = \int_0^x \frac{1}{\sqrt{1-t^2}}\,dt.$$

Begin with the series found in Exercise 2 that

$$\sqrt{1+x} = 1 + \frac{1}{2}x - \frac{1}{8}x^2 + \frac{1}{16}x^3 - \cdots + \frac{(-1)^{n-1}1\cdot 3\cdots(2n-3)}{2^n\cdot n!}x^n + \cdots.$$

Differentiate both sides, next multiply by 2, then replace x by $-t^2$, and finally integrate term-by-term to obtain the series

$$\arcsin x = x + \frac{1}{6}x^3 + \frac{3}{40}x^5 + \cdots + \frac{1\cdot 3\cdots(2n-1)}{2^n\cdot n!(2n+1)}x^{2n+1} + \cdots$$

$$= x + \sum_{n=1}^{\infty} \frac{1\cdot 3\cdots(2n-1)}{2^n\cdot n!(2n+1)}x^{2n+1} + \cdots.$$

(b) Use the ratio test to show that the interval of convergence for the series for $\arcsin x$ in part (a) is $-1 < x < +1$. Since the ratio test fails to apply at $x = \pm 1$, another test needs to be used for these values. Raabe's test can be used to establish convergence at the endpoints (see Problem 10).

7. **(a)** Substitute ix into the series for e^x and show that this series can be regrouped into the form $\cos x + i\sin x$.

(b) Use the result in part (a) to show that $e^{-ix} = \cos x - i\sin x$.

(c) Use the results in parts (a) and (b) to prove that

$$\cos x = \frac{e^{ix} + e^{-ix}}{2} \quad \text{and} \quad \sin x = \frac{e^{ix} - e^{-ix}}{2i}.$$

Problem 4. Evaluating $\int_0^1 x^k\,dx$ by Cavalieri Sums

Cavalieri was an Italian mathematician and a contemporary of the French mathematicians Descartes and Fermat. He was also a pupil of Galileo and some of his ideas about finding areas likely suggested the modern formula

$$\int x^n\,dx = \frac{x^{n+1}}{n+1}$$

to Isaac Newton. Cavalieri's method for approximating the area under the curve $f(x) = x^k$ over the interval [0, 1] was to divide the interval into n equal parts and then sum the areas of the n rectangles with height determined by the function value at the right endpoint of each subinterval. This approach is used in most calculus texts at the start of the chapter on integration and it gives the following upper bound for the area.

$$\sum_{i=1}^{n} f(x_i)\,\Delta x_i = \sum_{i=1}^{n} f\left(\frac{i}{n}\right)\cdot\frac{1}{n} = \frac{1}{n^{k+1}}\sum_{i=1}^{n} i^k$$

Cavalieri's next task was to find formulas for $\sum_{i=1}^{n} i^k$, and to assert that as n gets large, the resulting upper bound sums approach the desired value for the area. With significant effort, he obtained the formulas for $i = 1, 2, \ldots, 9$, which are given below.[1] While the first four formulas are familiar to today's calculus student, the others are likely new. One can only surmise that Cavalieri stopped at $k = 9$ because of understandable fatigue.

$$I_0 = \sum_{i=1}^{n} 1 = n$$

$$I_1 = \sum_{i=1}^{n} i = \frac{n}{2}(n+1)$$

$$I_2 = \sum_{i=1}^{n} i^2 = \frac{n}{6}(n+1)(2n+1)$$

$$I_3 = \sum_{i=1}^{n} i^3 = \frac{n^2}{4}(n+1)^2$$

$$I_4 = \sum_{i=1}^{n} i^4 = \frac{n}{30}(n+1)(2n+1)(3n^2+3n-1)$$

$$I_5 = \sum_{i=1}^{n} i^5 = \frac{n^2}{12}(n+1)^2(2n^2+2n-1)$$

$$I_6 = \sum_{i=1}^{n} i^6 = \frac{n}{42}(n+1)(2n+1)(3n^4+6n^3-3n+1)$$

$$I_7 = \sum_{i=1}^{n} i^7 = \frac{n^2}{24}(n+1)^2(3n^4+6n^3-n^2-4n+2)$$

$$I_8 = \sum_{i=1}^{n} i^8 = \frac{n}{90}(n+1)(2n+1)(5n^6+15n^5+5n^4-15n^3-n^2+9n-3)$$

$$I_9 = \sum_{i=1}^{n} i^9 = \frac{n^2}{20}(n+1)^2(2n^6+6n^5+n^4-8n^3+n^2+6n-3)$$

[1] See p. 129 in [1].

We show in Problem 28 that the Euler–Maclaurin formula provides an easier way to obtain these summation formulas. In Problem 16, we present the method of Fermat to find a value for $\int_0^1 x^k\,dx$ without using these summation formulas at all.

Example 1.

$$\begin{aligned}\int_0^1 x^2\,dx &= \lim_{n\to\infty}\sum_{i=1}^{n}\left(\frac{i}{n}\right)^2\cdot\frac{1}{n}\\ &= \lim_{n\to\infty}\frac{1}{n^3}\sum_{i=1}^{n} i^2\\ &= \lim_{n\to\infty}\frac{n(n+1)(2n+1)}{6n^3} = \frac{1}{3}.\end{aligned}$$

Exercises

1. These nine formulas from Cavalieri are recursively obtained. That is, to find the formula for $\sum_{i=1}^{n} i^4$, we need to know the formulas for

$$\sum_{i=1}^{n} i,\quad \sum_{i=1}^{n} i^2,\quad \text{and}\quad \sum_{i=1}^{n} i^3.$$

   Show how to obtain the above formula for I_4 by starting with the identity

$$(i+1)^5 - i^5 = 5i^4 + 10i^3 + 10i^2 + 5i + 1.$$

   See Example 5 in Problem 2.

2. Identify as many patterns as possible in the nine formulas given above. One example is that $n^2 | I_{2k+1}$ for all k. Then use these patterns to make conjectures about the formulas for I_{10} and I_{11}.

3. Show that the use of Cavalieri's method and formulas give

$$\int_0^1 x^4\,dx = \frac{1}{5}\quad\text{and}\quad\int_0^1 x^9\,dx = \frac{1}{10}.$$

4. See pp. 106–110 in [8] for a discussion of Faulhaber's formula, which generalizes and simplifies the above calculations. This selection also shows how the binomial theorem and Bernoulli numbers are connected in these summation formulas.

Section 2. Supplementary Problems

Problem 5. The Pervasive Nature of Sequences

Interestingly, most calculus texts do not formally introduce the concept of a sequence until the chapter in the middle of the book that is devoted to infinite series, even though several important sequences have already been presented by this time. Some of these earlier occurrences are described below.

1. Newton's method uses a recursively defined sequence $\{x_n\}$ to approximate the value of a zero of the equation $f(x) = 0$. We let x_1 be an integer value near the desired zero and define the subsequent approximations by

$$x_{n+1} = x_n - \frac{f'(x_n)}{f(x_n)}.$$

2. The sequence $\{e_n\}$ where

$$e_n = \left(1 + \frac{1}{n}\right)^n$$

is used to define the value for e, the base of the natural logarithm function.

3. The concept of a Riemann sum is introduced when defining the definite integral. A special case is obtained for a continuous function f by dividing the interval $[a, b]$ into n equal parts and determining n rectangles, each having one of these subintervals as its base and its height determined by the function value at the right endpoint of this subinterval. We then approximate the area under the curve $f(x)$ and above the interval $[a, b]$ by the sum of the areas of these n rectangles which is

$$S_n = \sum_{i=1}^{n} f(a + i\Delta x)\, \Delta x,$$

where

$$\Delta x = \frac{b - a}{n}.$$

The sequence $\{S_n\}$ converges to the value of the integral $\int_a^b f(x)\, dx$.

4. The trapezoid rule is used to approximate the value of a definite integral when we are unable to find an antiderivative for $f(x)$. We define the sequence of approximations by $\{x_n\}$ where

$$x_n = \frac{h}{2}\{f(a) + 2f(a + h) + 2f(a + 2h) + \cdots + 2f(a + (n - 1)h) + f(a + nh)\}$$

and

$$h = \frac{b - a}{n}.$$

5. The value for the improper integral

$$\int_1^\infty \frac{1}{x^p}\, dx$$

for $p > 1$ can be defined in terms of the limit of the sequence $\{x_n\}$ where

$$x_n = \int_1^n \frac{1}{x^p}\, dx.$$

Most real analysis texts present a rigorous unit on sequences near the beginning of the book. In this setting, several examples of sequences are presented to illustrate the theorems about sequences. Some of these examples are given below. None involves sequences of partial sums, which have an importance all their own in the setting of infinite series.

Example 1. Let $\{x_n\}$ be defined by

$$x_n = \sum_{i=1}^{2n} \frac{1}{i} - \sum_{i=1}^{n} \frac{1}{i} = \sum_{i=n+1}^{2n} \frac{1}{i}.$$

This sequence is monotone increasing because

$$\begin{aligned} x_{n+1} - x_n &= \left(\sum_{i=1}^{2n+2} \frac{1}{i} - \sum_{i=1}^{n+1} \frac{1}{i} \right) - \left(\sum_{i=1}^{2n} \frac{1}{i} - \sum_{1}^{n} \frac{1}{i} \right) \\ &= \frac{1}{2n+2} + \frac{1}{2n+1} - \frac{1}{n+1} \\ &= \frac{1}{2n+2} + \frac{1}{2n+1} - \frac{1}{2n+2} - \frac{1}{2n+2} \\ &= \frac{1}{(2n+1)(2n+2)} > 0. \end{aligned}$$

The sequence is bounded above because

$$x_n = \frac{1}{n+1} + \frac{1}{n+2} + \cdots + \frac{1}{2n} < \frac{n}{n+1} < 1,$$

and is therefore convergent.

We can also show that if $f(x) = 1/x$ is defined on the interval $[1, 2]$ and P_n is the partition of this interval into n equal parts, then the lower sum $L(f, P)$ is equal to x_n because

$$L(f, P) = \sum_{i=1}^{n} m_i \, \Delta x_i = \sum_{i=1}^{n} f\left(1 + \frac{i}{n}\right) \cdot \frac{1}{n} = \sum_{i=1}^{n} \frac{1}{1 + \frac{i}{n}} \cdot \frac{1}{n} = \sum_{i=1}^{n} \frac{1}{n+i} = x_n.$$

This observation suggests that $\{x_n\}$ converges to $\ln 2$.

Example 2. Let $\{x_n\}$ be defined by

$$x_n = \sum_{i=1}^{n} \frac{1}{i} - \ln n.$$

By considering areas under curves in a way similar to the method used in the proof of the integral test for series convergence, we can show this sequence is decreasing and bounded below, and therefore convergent. See Exercise 6. The value for this convergent sequence is known as Euler's constant and denoted by γ, which is the Greek letter gamma.[2] An air

[2] For an extensive discussion of γ, see Julian Havil, *Gamma: Exploring Euler's Constant*, Princeton University Press, Princeton, New Jersey, 2003.

of mystery surrounds this number because we still do not know whether it is rational or irrational.

Example 3. The sequence defined recursively by $x_n = \frac{1}{2}(x_{n-1} + x_{n-2})$ where $x_1 = 1$ and $x_2 = 2$ is interesting in that it is not monotone and bounded as are the sequences in Examples 1 and 2. However this sequence has monotone and bounded subsequences, which is the key to finding its limit. See Problem 22 for more discussion of this example and some extensions.

Example 4. We use the principle of mathematical induction to prove that the recursively defined sequence $\{x_n\}$ where

$$x_{n+1} = 3 - \frac{1}{x_n} \quad \text{and} \quad x_1 = 1$$

is increasing and bounded above.

We first state the proposition to be proved, namely

$$P(n) : 0 < x_n < x_{n+1},$$

and observe that $P(1)$ is true. After some algebraic steps, it follows that

$$3 - \frac{1}{x_n} < 3 - \frac{1}{x_{n+1}}$$

or $0 < x_{n+1} < x_{n+2}$, and so $P(n+1)$ is true.

We next form the new proposition

$$P(n) : 1 \le x_n \le 3.$$

$P(1)$ is trivially true. This inequality leads to

$$2 \le 3 - \frac{1}{x_n} \le \frac{8}{3},$$

so once again $P(n+1)$ is true.

Since $\{x_n\}$ is monotone and bounded, it must converge. If we let $\lim x_n = x = \lim x_{n+1}$, we have the equation

$$\lim_{n\to\infty} x_{n+1} = 3 - \frac{1}{\lim_{n\to\infty} x_n} \quad \text{or} \quad x = 3 - \frac{1}{x}.$$

This leads to the quadratic equation $x^2 - 3x + 1 = 0$, whose solution $(3 + \sqrt{5})/2$ satisfies the conditions of the given sequence. Since this value equals $2 + \tau$, where $\tau = (\sqrt{5} - 1)/2$ is the "golden ratio," it suggests the possibility that this sequence might be related to the Fibonacci sequence. Some details of this are contained in Exercise 7.

Exercises

1. Explain why we define the sequence $\{x_n\}$ in Newton's method by the formula

$$x_{n+1} = x_n - \frac{f(x_n)}{f'(x_n)}.$$

Hint: Consider the purpose of Newton's method.

2. Show that Newton's method yields

$$x_{n+1} = \frac{1}{2}\left(x_n + \frac{a}{x_n}\right)$$

as the recursive sequence that converges to $\sqrt{a}$.

3. Extend the work in Exercise 2 to find a recursively defined sequence that converges to $\sqrt[3]{a}$. Apply Newton's method to $f(x) = x^3 - a$.

4. Use the binomial expansion for $(a+b)^n$ to write out the expansion for

$$e_n = \left(1 + \frac{1}{n}\right)^n.$$

Then use this expansion to explain why $\{e_n\}$ is monotone increasing and bounded above.

5. **(a)** Extend the work done in Example 1 to show that the sequences $\{u_n\}$ and $\{v_n\}$ defined by

$$u_n = \sum_{i=1}^{3n} \frac{1}{i} - \sum_{i=1}^{n} \frac{1}{i} = \sum_{i=n+1}^{3n} \frac{1}{i} \quad \text{and} \quad v_n = \sum_{i=1}^{3n} \frac{1}{i} - \sum_{i=1}^{2n} \frac{1}{i}$$

are monotone increasing and bounded above, and therefore convergent.

(b) Show that $\{u_n\}$ converges to $\ln 3$ by considering a partition of the interval $[1, 3]$ into $2n$ equal parts and using the definition of the integral for $f(x) = 1/x$ on the interval $[1, 3]$ to show that the lower sum $L(f, P) = u_n$.

(c) In a similar way to the work in part (b), show that $\{v_n\}$ converges to $\ln 1.5$ by considering a partition of the interval $[2, 3]$ into n equal parts.

(d) Let

$$x_n = \sum_{i=1}^{tn} \frac{1}{i} - \sum_{i=1}^{rn} \frac{1}{i},$$

where r and t are positive integers so that $0 < r < t$. Prove that $\{x_n\}$ converges by showing it is monotone increasing and bounded above. Then show that $\{x_n\}$ converges to $\ln((t-r)/r)$ by considering a partition of the interval $[r, t]$ into $(t-r)n$ equal parts.

6. Verify that the sequence in Example 2 is monotone decreasing and bounded below.

Hint: Consider $\ln n$ as the area under the curve $f(x) = 1/x$ over the interval $[1, n]$, and consider

$$\sum_{i=1}^{n} \frac{1}{i}$$

as the sum of the area of n rectangles.

7. (a) Given the sequence

$$x_{n+1} = 3 - \frac{1}{x_n}$$

defined in Example 4, calculate the values for x_2, x_3, x_4, x_5 and x_6, expressing your answers as rational fractions.

(b) Observe that the values found in part (a) suggest the conjecture that

$$x_n = \frac{F_{2n-2}}{F_{2n-4}},$$

where $\{F_n\}$ denotes the Fibonacci sequence which begins with the terms $\{1, 2, 3, 5, 8, 13, 21, 34, 55, 89, 144, \ldots\}$.

(c) Verify the identity that $F_{n-2} + F_{n+2} = 3F_n$, using the definition that $F_n = F_{n-1} + F_{n-2}$.

(d) Explain how the identity in part (c) applies to the sequence in part (a).

Problem 6. Counterexamples

The study of examples and counterexamples is a very useful tool to sharpen one's understanding of a definition or theorem. The following exercises help us to distinguish between various concepts in analysis. These occur throughout an analysis course, and are collected together here to provide a review as well as to reinforce the importance of such examples in analysis. Provide at least one example for each item given below. For a more complete list, see the book described below.[3]

1. A function that is continuous at a point but not differentiable at that point.
2. A function that is differentiable at a point, but so that the derivative is not continuous at that point.
3. A function that is discontinuous at every point of its domain.
4. A function that is integrable on $[a, b]$, but not continuous on $[a, b]$.
5. A function that is integrable on $[a, b]$ even though it has infinitely many discontinuities on $[a, b]$.
6. A function that is bounded, but not integrable on $[a, b]$.

[3] Bernard R. Gelbaum and John M.H. Olmsted, *Counterexamples in Analysis*, Holden-Day, Inc., San Francisco, 1964. This book is in print again as a 2003 Dover paperback.

7. A function $f(x)$ so that $f^{(n)}(0)$ exists for all n, but so that $f(x)$ has no Maclaurin series representation. See Problem 15.
8. A function $f(x)$ for which we know an expression for the general term of its Maclaurin series even though we cannot directly calculate the expression for $f^{(n)}(x)$. See Problem 3.
9. A function that is continuous on a set S without being uniformly continuous there.
10. A function that has an unbounded derivative on S, but is still uniformly continuous there.
11. A divergent series $\sum a_n$ where $\lim a_n = 0$, but $\lim_{n\to\infty} na_n \neq 0$.
12. A divergent series $\sum a_n$ where $\lim na_n = 0$.
13. A sequence of continuous functions $\{f_n(x)\}$ on S that converge to a discontinuous function $f(x)$ on the set S.
14. A function that is Lebesgue integrable, even though it is not Riemann integrable.
15. A set that is uncountable and yet has Lebesgue measure zero.
16. A divergent sequence that has a convergent subsequence.

Problem 7. Some Examples of Unusual Functions

The concept of a function and its properties of continuity and differentiability have evolved slowly over time. Euler thought a function was continuous only if it could be expressed by a single formula. For example, the function

$$f(x) = \begin{cases} x & \text{if } x \geq 0 \\ -x & \text{if } x < 0 \end{cases}$$

would have been considered discontinuous by Euler in the 18th century, but he would consider the same function to be continuous if written in modern notation using a single expression as $f(x) = |x|$.

Beginning in the 19th century, some unusual examples of functions appeared which helped to clarify the basic definitions and to show that intuition cannot always be trusted. Until then mathematicians believed there was no real difference between continuity and differentiability, and that most functions were basically continuous except for a few values. Several of these examples are given below.

Example 1. The following function was introduced by Dirichlet in 1829.

$$f(x) = \begin{cases} a & \text{if } x \text{ is rational} \\ b & \text{if } x \text{ is irrational} \end{cases}$$

It is an interesting function because it is discontinuous at every real number, assuming that $a \neq b$. This can be verified by use of the sequence form of the definition of continuity. By the density property of the irrational numbers, choose a sequence $\{t_n\}$ of irrationals that converges to a given rational number r. The sequence $\{f(t_n)\} = \{b\}$ does not converge to $f(r) = a$, so the function f must be discontinuous at every rational number. A similar argument can be applied to an arbitrary irrational number. This function is also an example of a bounded function that is not integrable in the Riemann sense, although it is Lebesgue integrable.

Example 2.

$$f(x) = \frac{1}{[x]} \quad \text{and} \quad g(x) = \left[\frac{1}{x}\right],$$

where $[x]$ denotes the greatest integer n so that $n \leq x$, are two examples of functions that have an infinite number of jump discontinuities. The function $g(x)$ has all its discontinuities in the bounded interval $[-1, 1]$ and has an infinite discontinuity there as well.

Example 3. The following function is usually attributed to K.J. Thomae, who presented it in 1875, and is an example of a function continuous for every irrational number, but discontinuous for every rational number.

$$f(x) = \begin{cases} \frac{1}{n} & \text{if } x \text{ is rational of the form } x = \frac{m}{n} \text{ where } (m, n) = 1 \\ 0 & \text{if } x \text{ is irrational} \end{cases}$$

On the other hand, it is not possible to find a function that is continuous on the rationals but discontinuous on the irrationals. Bill Dunham wrote about a proof of this fact by Vito Volterra in 1881 (see [**9**]).

Example 4. The function defined below has the property that $f^{(n)}(0)$ exists, but $f^{(n)}(x)$ is not continuous at $x = 0$ and $f^{(k)}(0)$ does not exist for $k > n$.

$$f(x) = \begin{cases} x^{2n} \sin \frac{1}{x} & \text{if } x \neq 0 \\ 0 & \text{if } x = 0 \end{cases}$$

For example, when $n = 1$ and

$$f(x) = x^2 \sin \frac{1}{x}$$

if $x \neq 0$ and $f(0) = 0$, it follows that $f'(0)$ exists but $f'(x)$ is not continuous at $x = 0$. We know that $f'(0) = 0$ because

$$f'(0) = \lim_{x \to 0} \frac{f(x) - f(0)}{x - 0} = \lim_{x \to 0} x \sin\left(\frac{1}{x}\right) = 0.$$

Since

$$f'(x) = 2x \sin \frac{1}{x} - \cos \frac{1}{x},$$

$f'(x)$ is not continuous at $x = 0$ because

$$\lim_{x\to 0} \cos\frac{1}{x}$$

does not exist.

Example 5. If you have the impression from the above examples that it is only the discontinuous functions that possess unusual properties, then you may be surprised by an example Weierstrass presented of a function that is continuous everywhere on the reals but does not have a derivative at any point.[4] This very non-intuitive example presented in the form of a trigonometric series was considered a revolutionary idea by most mathematicians in the last part of the 19th century. The concept is perhaps easier to accept today because we are more familiar with fractals which exhibit this property. Weierstrass's famous example is defined by

$$f(x) = \sum_{n=0}^{\infty} b^n \cos(a^n \pi x)$$

where a is an odd integer, $0 < b < 1$ and $ab > 1 + 3\pi/2$.

Exercises

1. Prove that the function in Example 1 is discontinuous at each irrational.
2. Graph the two functions that are defined in Example 2. Also give the domain of definition for each function and list all the points of discontinuity.
3. Prove that the function defined in Example 3 is discontinuous at each rational by using the sequence form of continuity.
4. Prove that the function defined in Example 3 is continuous at each irrational. Hint: Given $\epsilon > 0$, how many rational numbers r satisfy $f(r) \geq \epsilon$?
5. Verify that the following function satisfies the properties that $f''(0)$ exists, but $f''(x)$ is not continuous at $x = 0$.

$$f(x) = \begin{cases} x^4 \sin\frac{1}{x} & \text{if } x \neq 0 \\ 0 & \text{if } x = 0 \end{cases}$$

Problem 8. Identifying Interior, Exterior, Boundary, and Limit Points

We describe various categories of points for a set in Essay 10 on sets and topology. The purpose of this problem is to apply these definitions to several subsets of the set $\mathfrak{R}$ of real numbers. For each set S below, fill in the chart identifying int S (the set of all interior points

[4]See pp. 259–63 in [5] for the historical context of this example and a proof of the details.

of S), ext S (the set of all exterior points of S), bd S (the set of all boundary points of S), and S' (the set of all limit points of S). Observe that the non-empty sets of int S, ext S, and bd S form a partition of $\Re$. Is there a relationship between the sets bd S and S'?

		int S	ext S	bd S	S'
(1)	$S = (a, b)$				
(2)	$S = [a, b]$				
(3)	$S = [a, \infty)$				
(4)	$S = (-\infty, b)$				
(5)	$S = \mathbf{N}$				
(6)	$S = \mathbf{Q}$ (rationals)				
(7)	$S = \Re$				
(8)	$S = \left\{\frac{1}{n} : n \in \mathbf{N}\right\}$				
(9)	$S = \{\sin r : r \in \mathbf{Q} \text{ and } 0 \leq r < \pi\}$				
(10)	$S = \{\sin x : x \in \Re \text{ and } 0 \leq x < \pi\}$				
(11)	$S = \{\sin n : n \in \mathbf{N}\}$				
(12)	$S = \left\{\tan r : r \in \mathbf{Q} \text{ and } 0 < r < \frac{\pi}{2}\right\}$				

Problem 9. Exploring Uniform Continuity

Uniform continuity is one of the more difficult concepts in analysis. You can enhance your understanding of this concept by thinking of connections with previous concepts and results. Intuition and guesswork are important tools at this point. Proposed theorems should be checked by seeing if they hold for several examples. If a counterexample is found by this process, the desired theorem cannot be true, but it may be the statement of the result can be altered somewhat to avoid the conflict seen in the counterexample.

You are now given the opportunity to practice this approach. Six possible conjectures about uniform continuity are given in the chart below. All of them should seem plausible. It turns out that whether the domain set S is compact or not makes a difference, so each conjecture is to be checked in three different settings—when S is closed and bounded, and hence compact, when S is bounded but not closed, and when S is closed but not bounded. In each of the 18 boxes, decide whether the result is always true (provide a proof or a reason) or whether the result is sometimes false (provide a counterexample). Conjecture 1 is done as an example.

	$S = [a, b]$ is closed and bounded	$S = (a, b)$ is bd. but not closed	$S = [a, \infty)$ is closed but not bounded
Conjecture 1. If f is continuous on S, then f is uniformly continuous on S.	TRUE by a standard theorem	FALSE $f(x) = 1/x$ on $S = (0, 1)$	FALSE $f(x) = x^2$ on $S = [0, \infty)$
Conjecture 2. If f is bounded and cont. on S, then f is uniformly continuous on S.			
Conjecture 3. If f is unbounded on S, then f is not uniformly continuous on S.			
Conjecture 4. If f' is unbounded on S, then f is not uniformly continuous on S.			
Conjecture 5. If f' is bounded on S, then f is uniformly continuous on S.			
Conjecture 6. If f and f' are unbounded on S, then f is not uniformly cont. on S.			

If you need or want some hints for this problem, the following comments will likely be of help.

The following five results can be used to support the assignment of TRUE in appropriate places in the chart. Provide proofs for Theorems A, B, and C.

1. Theorem A: If f is continuous on the compact set $[a, b]$, then f is uniformly continuous there.

2. Theorem B: If f is uniformly continuous on (a, b), then it is possible to extend f to a function that is continuous on $[a, b]$. This can be proven by use of the Bolzano–Weierstrass theorem for sequences.
3. Theorem C: If f has a bounded derivative on an interval, then f is uniformly continuous on that interval. This can be proven as a corollary to the mean value theorem.
4. The contrapositive statement D: $(p \Rightarrow q) \Longleftrightarrow (\neg q \Rightarrow \neg p)$
5. A rule of logic E: $(p \Rightarrow q) \Longrightarrow ((p \wedge r) \Rightarrow q)$

Each of the following functions provides a possible counterexample to support the assignment of FALSE in one or more of the boxes in the chart.

1. $f_1(x) = 1/x$ on $(0, 1)$
2. $f_2(x) = x^2$ on $[1, \infty)$
3. $f_3(x) = \sin(1/x)$ on $(0, 1)$
4. $f_4(x) = \sin(x^2)$ on $[1, \infty)$
5. $f_5(x) = x$
6. $f_6(x) = \sqrt{x}$

Problem 10. Comparing the Ratio Test, the Root Test, and Raabe's Test

Of these three tests, the ratio test is likely the only one familiar to calculus students. This test provides the best method to find the interval of convergence for a power series. The root test does not easily apply to most of the familiar series considered in calculus. Raabe's test is a very special test and only applies to series for which the ratio test fails.

We illustrate by applying the ratio test and the root test to the series

$$\sum_{n=1}^{\infty} a_n = \sum_{n=1}^{\infty} \frac{n!}{n^n}.$$

Since

$$\frac{a_{n+1}}{a_n} = \frac{(n+1)!}{(n+1)^{(n+1)}} \cdot \frac{n^n}{n!} = \left(\frac{n}{n+1}\right)^n = \frac{1}{(1+\frac{1}{n})^n}$$

it follows that

$$\lim_{n\to\infty} \frac{a_{n+1}}{a_n} = \frac{1}{e} < 1$$

and so

$$\sum_{n=1}^{\infty} \frac{n!}{n^n}$$

converges by the ratio test. On the other hand, we are unable to apply the root test because of the difficulty in evaluating

$$\lim_{n\to\infty} \frac{\sqrt[n]{n!}}{n}$$

(see Exercises 1 and 2). These results seem to indicate that the ratio test is more general than the root test, but the opposite is true. This fact is based on the key inequality that

$$\underline{\lim}\, \frac{a_{n+1}}{a_n} \le \underline{\lim}\, \sqrt[n]{a_n} \le \overline{\lim}\, \sqrt[n]{a_n} \le \overline{\lim}\, \frac{a_{n+1}}{a_n} \tag{1}$$

where the *limit superior* of a_n (denoted by $\overline{\lim}\, a_n$ or $\limsup a_n$) and the *limit inferior* of a_n (denoted by $\underline{\lim}\, a_n$ or $\liminf a_n$) are defined as follows.

If the sequence $\{a_n\}$ has no upper bound, then $\overline{\lim}\, a_n = +\infty$. If there is some L so that for every $\epsilon > 0$, $L - \epsilon < a_n$ holds for infinitely many values of n and $a_n < L + \epsilon$ holds for all but a finite number of n, then $\overline{\lim}\, a_n = L$. Otherwise, $\overline{\lim}\, a_n = -\infty$. Another way to express this concept is to say that $\overline{\lim}\, a_n$ is the supremum of all the limit points of $\{a_n\}$. The $\underline{\lim}\, a_n$ concept can be defined in a similar way (see Exercise 5). An outline for a proof of inequality (1) can be found in the literature.[5]

From this inequality we know that if

$$\lim_{n\to\infty} \frac{a_{n+1}}{a_n}$$

exists and equals L, then $\lim_{n\to\infty} \sqrt[n]{a_n}$ also exists and equals L. So if the ratio test applies to assert the convergence of a given series $\sum a_n$, so must the root test.

The next thing to observe is that the ratio and root tests can be stated in a more general form than is usually found in a calculus text, where the concepts of $\underline{\lim}\, a_n$ and $\overline{\lim}\, a_n$ are not introduced. We assume that all $a_n > 0$.

Generalized form of the ratio test.

$$\text{If } \overline{\lim}\, \frac{a_{n+1}}{a_n} < 1, \quad \text{then } \sum_{n=1}^{\infty} a_n \text{ converges.}$$

$$\text{If } \underline{\lim}\, \frac{a_{n+1}}{a_n} > 1, \quad \text{then } \sum_{n=1}^{\infty} a_n \text{ diverges.}$$

Otherwise the test fails.

Generalized form of the root test.

$$\text{If } \overline{\lim}\, \sqrt[n]{a_n} < 1, \quad \text{then } \sum_{n=1}^{\infty} a_n \text{ converges.}$$

---

[5] James Kirkwood, *An Introduction to Analysis*, PWS-Kent Publishing Company, Boston, 1989, p. 187.

$$\text{If } \underline{\lim}\, \sqrt[n]{a_n} > 1, \quad \text{then} \quad \sum_{n=1}^{\infty} a_n \text{ diverges.}$$

Otherwise the test fails.

The key inequality (1) above again shows that if $\sum_{n=1}^{\infty} a_n$ converges by the generalized ratio test, then it must converge by the generalized root test also. Whether

$$\overline{\lim}\, \frac{a_{n+1}}{a_n} \quad \text{and} \quad \overline{\lim}\, \sqrt[n]{a_n}$$

can be evaluated for a specific series $\sum a_n$ is another matter.

Example 1. The series

$$a + b + a^2 + b^2 + a^3 + b^3 + \cdots \quad \text{where} \quad 0 < a < b < 1$$

is an example of a series for which the generalized ratio test fails, but the generalized root test implies convergence. The generalized ratio test fails because the expression

$$\frac{a_{n+1}}{a_n}$$

has either the form

$$\frac{a^k}{b^{k-1}} = a \cdot \left(\frac{a}{b}\right)^{k-1} \quad \text{or} \quad \frac{b^k}{a^k} = \left(\frac{b}{a}\right)^k,$$

which means that

$$\underline{\lim}\, \frac{a_{n+1}}{a_n} = 0 \quad \text{and} \quad \overline{\lim}\, \frac{a_{n+1}}{a_n} = +\infty > 1.$$

The root test implies convergence because $\sqrt[n]{a_n}$ has either the form

$$(a^k)^{1/2k-1} \quad \text{or} \quad (b^k)^{1/2k} = \sqrt{b}.$$

Therefore, $\overline{\lim}\, \sqrt[n]{a_n} = \sqrt{b} < 1$. Since $\underline{\lim}\, \sqrt[n]{a_n} = \sqrt{a} < \sqrt{b}$, the usual form of the root test fails to apply to this series.

This is a somewhat artificial example, as this series can be handled by use of geometric series, observing that

$$\begin{aligned} S_{2k} &= (a + b + a^2 + b^2 + \cdots + a^k + b^k) \\ &= (a + a^2 + a^3 + \cdots + a^k) + (b + b^2 + b^3 + \cdots + b^k) \\ &= \frac{a - a^{k+1}}{1-a} + \frac{b - b^{k+1}}{1-b} \end{aligned}$$

and so the series converges to

$$\lim_{k\to\infty} S_{2k} = \frac{a}{1-a} + \frac{b}{1-b}.$$

We know the ratio test fails to apply at the endpoints of the interval of convergence for power series. While other tests such as the alternating series test usually apply at these endpoints, the series of constants for such important Maclaurin series as those for

$$f(x) = \arcsin x \quad \text{and} \quad f(x) = (1+x)^k$$

require a more complicated test. The usual choice is Raabe's test, which was established by Josef Ludwig Raabe in 1832. It is stated and proved below, and then applied to these important series in the examples and exercises.

Raabe's test. Assume that

$$\lim_{n\to\infty}\left|\frac{a_{n+1}}{a_n}\right| = 1 \quad \text{and} \quad \lim_{n\to\infty} n\left(1 - \left|\frac{a_{n+1}}{a_n}\right|\right) = t.$$

Then $\sum_{n=1}^{\infty} a_n$ converges absolutely if $t > 1$. Also, $\sum_{n=1}^{\infty} a_n$ diverges if $t < 1$.

Proof We begin by observing two facts. First, for $p > 1$, the p-series

$$\sum c_n = \sum \frac{1}{n^p}$$

converges and

$$\frac{c_{n+1}}{c_n} = \frac{\frac{1}{(n+1)^p}}{\frac{1}{n^p}} = \left(\frac{n}{n+1}\right)^p = \left(\frac{n+1}{n}\right)^{-p} = \left(1+\frac{1}{n}\right)^{-p}.$$

The second fact is that when Taylor's theorem with remainder is applied to $f(x) = (1+x)^{-p}$ using $a = 0$ and $n = 1$, we obtain

$$(1+x)^{-p} = 1 - px + \frac{p(p+1)}{2(1+c)^{p+2}}x^2 \quad \text{for some } c \in (0, x).$$

So if $x = 1/n$, this becomes

$$\begin{aligned}\left(1+\frac{1}{n}\right)^{-p} &= 1 - \frac{p}{n} + \frac{p(p+1)}{2(1+c)^{p+2}}\frac{1}{n^2} \quad \text{for some } c \text{ so that } 0 < c < \frac{1}{n} \\ &> 1 - \frac{p}{n}.\end{aligned}$$

Since

$$\lim_{n\to\infty} n\left(1 - \left|\frac{a_{n+1}}{a_n}\right|\right) = t > 1,$$

there is some value $p > 1$ so that

$$n\left(1 - \left|\frac{a_{n+1}}{a_n}\right|\right) > p \quad \text{for all } n > \text{some } N$$

which is equivalent to

$$\left|\frac{a_{n+1}}{a_n}\right| < 1 - \frac{p}{n}.$$

Putting these results together, we have for $n > N$,

$$\left|\frac{a_{n+1}}{a_n}\right| < 1 - \frac{p}{n} < \left(1 + \frac{1}{n}\right)^{-p} = \frac{c_{n+1}}{c_n}$$

which implies that the given series $\sum_{n=1}^{\infty} a_n$ converges absolutely by use of the ratio form of the comparison test (see Theorem 4 in Essay 8). ■

Example 2. In Exercise 2 of Problem 3, we showed that for $-1 < x < 1$,

$$\sqrt{x+1} = 1 + \frac{1}{2}x + \sum_{n=2}^{\infty} \frac{(-1)^{n-1} 1 \cdot 3 \cdots (2n-3)}{2^n \cdot n!} x^n$$

and that the ratio test fails when $x = \pm 1$. When $x = -1$, we get the series

$$1 - \frac{1}{2} - \sum_{n=2}^{\infty} \frac{1 \cdot 3 \cdots (2n-3)}{2^n n!}.$$

For this series,

$$\left|\frac{a_{n+1}}{a_n}\right| = \frac{2n-1}{2n+2}$$

and so

$$n\left(1 - \left|\frac{a_{n+1}}{a_n}\right|\right) = n\left(1 - 2n - \frac{1}{2n+2}\right) = \frac{3n}{2n+2}.$$

Since

$$\left\{\frac{3n}{2n+2}\right\}$$

converges to $t = \frac{3}{2} > 1$, this series converges absolutely. The same conclusion follows if $x = +1$.

Exercises

1. (a) Use the root test to give a simple proof of the following result. If

$$\lim_{n \to \infty} \sqrt[n]{a_n} = 0,$$

then $\lim_{n \to \infty} a_n = 0$, assuming that all $a_n > 0$.

(b) Give an example to show that the converse of the result in part (a) is not true. So even though

$$\lim_{n\to\infty} \frac{n!}{n^n} = 0,$$

it does not imply

$$\lim_{n\to\infty} \sqrt[n]{\frac{n!}{n^n}} = 0$$

(see Exercise 2).

2. Prove that if $\lim_{n\to\infty} a_n = 0$, then if $\lim_{n\to\infty} \sqrt[n]{a_n}$ exists, the value must be ≤ 1. Note that this result is not strong enough to test the series

$$\sum_{n=1}^{\infty} \frac{n!}{n^n}$$

by the root test.

3. (a) Use the ratio test to show that the Maclaurin series expansion for $\arcsin x$ given below converges for $-1 < x < 1$.

$$\arcsin x = x + \sum_{n=1}^{\infty} \frac{1 \cdot 3 \cdots (2n-1)}{2^n \cdot n!(2n+1)} x^{2n+1}$$

(b) Use Raabe's test to show that the series in part (a) converges for $x = 1$. Therefore this series converges absolutely and uniformly on the interval $[-1, 1]$.

4. (a) Show that the Maclaurin series expansion for $f(x) = (1+x)^k$ is as follows.

$$(1+x)^k = 1 + \sum_{n=1}^{\infty} \frac{k(k-1)(k-2)\cdots(k-n+1)}{n!} x^n$$

(b) Use the ratio test to show that the series in part (a) converges when $-1 < x < 1$.

(c) Use Raabe's test to show that the series in part (a) converges when $x = \pm 1$.

5. Give a definition for the limit inferior of $\{a_n\}$.

Problem 11. Various Proofs that $\sum_{n=1}^{\infty} \frac{1}{n^2} = \frac{\pi^2}{6}$

One of the most challenging problems in the history of calculus has been the search for methods to find the exact value of a convergent infinite series. We summarize some of the more successful methods.

1. The *geometric series* result that

$$\sum_{n=0}^{\infty} r^n = \frac{1}{1-r} \quad \text{if} \quad |r| < 1$$

is perhaps the earliest general method, dating back to the time of the Greeks when Zeno showed geometrically that $1 = \frac{1}{2} + \frac{1}{4} + \frac{1}{8} + \cdots$ and Archimedes used the formula $1 + \frac{1}{4} + \frac{1}{16} + \frac{1}{64} + \cdots = \frac{4}{3}$ to find the area of a parabolic segment.[6]

2. The *telescoping principle* was used by Leibniz to sum the reciprocals of the triangular numbers. This method enables us to find exact values for series such as

$$\sum_{n=2}^{\infty} \frac{1}{n^2-1}, \quad \sum_{n=1}^{\infty} \frac{1}{n^2+n}, \quad \text{and} \quad \sum_{n=2}^{\infty} \frac{1}{n^3-n}$$

(see also Problem 2).

3. Substituting values for x in the *Maclaurin series* representation of a function is another fruitful method. Some well-known examples are to let $x = 1$ in the series

$$\arctan x = x - \tfrac{1}{3}x^3 + \tfrac{1}{5}x^5 - \cdots$$

to get

$$\frac{\pi}{4} = 1 - \frac{1}{3} + \frac{1}{5} - \frac{1}{7} + \cdots$$

or to let $x = 1$ in the series

$$e^x = 1 + x + \tfrac{1}{2!}x^2 + \tfrac{1}{3!}x^3 + \cdots$$

to get

$$e = 1 + 1 + \tfrac{1}{2!} + \tfrac{1}{3!} + \cdots.$$

However, none of these methods applies to the famous problem of finding the exact value for the p-series

$$\sum_{n=1}^{\infty} \frac{1}{n^2}.$$

This problem was one of many proposed by James and John Bernoulli at the end of the 17th century when calculus was in its infancy as a formal discipline. Although the Bernoulli brothers or Newton or Leibniz solved many problems, such as the brachistochrone problem,[7] the exact sum for the series

$$\sum_{n=1}^{\infty} \frac{1}{n^2}$$

eluded everyone until Euler made his presence felt on the mathematical landscape with his unique solution in 1735. Before this success, even Euler had found only methods to approximate this series value with great accuracy, but without the desired "exact" value.[8]

---

[6] See pp. 108–115 in **[27]** or pp. 233–6 in **[28]**.
[7] See pp. 308–17 in **[28]**.
[8] See pp. 43–6 in **[10]**.

Example 1. Euler's solution was based upon the Maclaurin series for $\sin x$ and an algebraic result about the sum of the reciprocals of the zeroes of a polynomial. It is not difficult to show that if $r_1, r_2, \ldots, r_n$ are the n zeroes of the polynomial $1 + a_1 x + a_2 x^2 + \cdots + a_n x^n$, then

$$\sum_{i=1}^{n} \frac{1}{r_i} = -a_1$$

(see Exercises 1 and 2). Euler assumed this result would also be true for an infinite polynomial. Since $\sin x = x - \frac{1}{3!}x^3 + \frac{1}{5!}x^5 - \cdots$ and since $\sin x$ has $0, \pm\pi, \pm 2\pi, \pm 3\pi, \ldots$ as zeroes, it follows that $1 - \frac{1}{3!}x^2 + \frac{1}{5!}x^4 - \cdots$ will have $\pm\pi, \pm 2\pi, \pm 3\pi, \ldots$ as zeroes, or $1 - \frac{1}{3!}y + \frac{1}{5!}y^2 - \cdots$ will have $\pi^2, 4\pi^2, 9\pi^2, \ldots$ as zeroes. Therefore

$$\sum_{n=1}^{\infty} \frac{1}{n^2 \pi^2} = -\left(-\frac{1}{3!}\right),$$

from which it follows that

$$\sum_{n=1}^{\infty} \frac{1}{n^2} = \frac{\pi^2}{6}.$$

Use of the Fourier series expansion for $f(x) = x^2$ is another means for getting this same formula. The details are in Exercises 2 and 3 from Problem 31. Bill Dunham discusses an alternative method[9] that Euler used to prove that

$$\sum_{n=1}^{\infty} \frac{1}{n^2} = \frac{\pi^2}{6},$$

and refers to a recent article (see **[24]**) where several modern and more rigorous solutions are discussed. Additional proofs are in the literature.[10]

Exercises

1. If the quadratic polynomial $x^2 + a_1 x + a_0$ has r_1 and r_2 as zeroes, then

$$x^2 + a_1 x + a_0 = (x - r_1)(x - r_2) = x^2 - (r_1 + r_2)x + r_1 r_2$$

implies that

$$r_1 + r_2 = -a_1 \quad \text{and} \quad r_1 r_2 = a_0,$$

so

$$\frac{1}{r_1} + \frac{1}{r_2} = \frac{r_1 + r_2}{r_1 r_2} = \frac{-a_1}{a_0}.$$

---

[9]See pp. 55–7 in **[10]**.

[10]See pp. 270–1 and 323–5 in **[28]**.

If r_1, r_2, and r_3 are the zeroes of the cubic polynomial $x^3 + a_2x^2 + a_1x + a_0$, show that

$$\frac{1}{r_1} + \frac{1}{r_2} + \frac{1}{r_3} = \frac{-a_1}{a_0}.$$

2. If $r_1, r_2, \ldots, r_n$ are the zeroes of $1 + a_1x + a_2x^2 + \cdots + a_nx^n$, then by the factor theorem

$$1 + a_1x + a_2x^2 + \cdots + a_nx^n = a_n(x - r_1)(x - r_2)\cdots(x - r_n)$$

and

$$a_n(-1)^n r_1 r_2 \cdots r_n = 1$$

by equating the constant term in both sides of the above equation. Use this information to prove that

$$1 + a_1x + a_2x^2 + \cdots + a_nx^n = \left(1 - \frac{x}{r_1}\right)\left(1 - \frac{x}{r_2}\right)\cdots\left(1 - \frac{x}{r_n}\right)$$

from which we find by equating coefficients of the x term that

$$\sum_{i=1}^{n} \frac{1}{r_i} = -a_1.$$

3. (a) Apply Euler's argument given in Example 1 to the zeroes of

$$\cos x = 1 - \frac{x^2}{2!} + \frac{x^4}{4!} - + \cdots$$

in order to establish the formula

$$1 + \frac{1}{3^2} + \frac{1}{5^2} + \cdots = \frac{\pi^2}{8}.$$

(b) Use the result in part (a) to prove that

$$\sum_{n=1}^{\infty} \frac{1}{n^2} = \frac{\pi^2}{6}.$$

4. Another general series that Jacob Bernoulli solved was

$$\frac{a}{b} + \frac{a+c}{bd} + \frac{a+2c}{bd^2} + \frac{a+3c}{bd^3} + \cdots = \frac{d(ad - a + c)}{b(d^2 - 2d + 1)}.$$

His method was to rewrite this series, where the numerators form an arithmetic sequence and the denominators form a geometric sequence, as an infinite number of geometric series whose sum is another geometric series.[11]

[11] For additional help with this problem, see pp. 40–1 in [**10**].

Problem 12. Finding a Value for $\sum_{n=1}^{\infty} \frac{1}{n^3}$

One of the most remarkable discoveries in the historical development of calculus is Euler's result from the 1730s that

$$\sum_{n=1}^{\infty} \frac{1}{n^2} = \frac{\pi^2}{6}$$

(see Problem 11). Perhaps even more remarkable was Euler's persistence to extend this result to the formulas

$$\sum_{n=1}^{\infty} \frac{1}{n^4} = \frac{\pi^4}{90}, \quad \sum_{n=1}^{\infty} \frac{1}{n^6} = \frac{\pi^6}{945},$$

and similar formulas for every series of the form

$$\sum_{n=1}^{\infty} \frac{1}{n^{2k}} \quad \text{for } k = 1, 2, \ldots, 13$$

(see Problem 28). However, he was unable to obtain a similar formula for the series

$$\sum_{n=1}^{\infty} \frac{1}{n^3}.$$

Even though many alternate methods have been found over the years to duplicate Euler's results for the value of even exponents of the p-series, finding a closed form expression for the $p = 3$ series remains an unsolved problem to the present day.[12] The exercises present various methods for approximating the value of

$$S = \sum_{n=1}^{\infty} \frac{1}{n^3}.$$

Exercises

1. The integral test provides the inequality

$$S_{k-1} + \int_k^{\infty} \frac{1}{x^3}\,dx < S < S_k + \int_k^{\infty} \frac{1}{x^3}\,dx$$

where

$$S_k = \sum_{i=1}^{k} \frac{1}{i^3}.$$

[12] See pp. 59–60 in [**10**].

This provides upper and lower approximations for S with an error less than $a_k = 1/k^3$ (see Theorem 6 in Essay 8). Use this result to find an approximation for S with error less than 10^{-2}, and also for an approximation with error less than 10^{-4}.

2. The value for the alternating series

$$T = \sum_{n=1}^{\infty} (-1)^{n-1} \frac{1}{n^3}$$

can be approximated by

$$T_k = \sum_{i=1}^{k} (-1)^{i-1} \frac{1}{i^3}$$

with an error that is less than

$$a_{k+1} = \frac{1}{(k+1)^3}$$

(see Theorem 8 in Essay 8). By writing the expressions for S and T and forming $S - T$, it is possible to show that $S = \frac{4}{3}T$ (see Problem 2). Therefore, we can use an approximation for T to find an approximation for S. Use this idea to approximate S with an error less than 10^{-4}, and compare with your result in Exercise 1.

3. In the 1730s during his first stay at the St. Petersburg Academy in Russia, Euler developed the result that is known today as the Euler–Maclaurin formula. This formula is found in the Euler–Stirling letters contained in Selection 2 in Part IV. Use this formula (applied to $f(x) = x^3$) to obtain an approximation for the value of S by following Euler's example with $f(x) = x^2$. Is there a way to measure the error in using this approximation?

4. Show the Fourier series expansion for the function $f(x) = x^3$ does not give a closed form expression for the series

$$\sum_{n=1}^{\infty} \frac{1}{n^3}$$

even though the Fourier series expansion for $f(x) = x^2$ does lead to the formula

$$\sum_{n=1}^{\infty} \frac{1}{n^2} = \frac{\pi^2}{6}.$$

See Example 2 in Problem 31.

Problem 13. Rearrangements of Conditionally Convergent Series

A *series of mixed signs* refers to a series with an infinite number of both positive and negative terms. A *rearrangement* of such a series is determined by using the same terms, but changing the order in which an infinite number of these terms occur. So

$$1 + \tfrac{1}{3} - \tfrac{1}{2} + \tfrac{1}{5} + \tfrac{1}{7} - \tfrac{1}{4} + + - \cdots$$

is a rearrangement of

$$1 - \tfrac{1}{2} + \tfrac{1}{3} - \tfrac{1}{4} + - \cdots,$$

while

$$1 + \tfrac{1}{2} - \tfrac{1}{3} + \tfrac{1}{4} + \tfrac{1}{5} - \tfrac{1}{6} + + - \cdots$$

is not. If the series $\sum_{n=1}^{\infty} |a_n|$ converges, there is a theorem asserting that the series of mixed signs $\sum_{n=1}^{\infty} a_n$ also converges. This is called *absolute convergence*. One consequence of absolute convergence is that every rearrangement of $\sum_{n=1}^{\infty} a_n$ also converges and to the same value to which $\sum_{n=1}^{\infty} a_n$ converges. On the other hand, if $\sum_{n=1}^{\infty} a_n$ converges, but $\sum_{n=1}^{\infty} |a_n|$ diverges, then $\sum_{n=1}^{\infty} a_n$ is said to possess *conditional convergence*. There is no general result about the convergence of rearrangements of conditionally convergent series. Each one has to be handled separately. We consider several examples of this after presenting a generalization of the alternating series test.

The basic idea underlying the proof for the alternating series test (see Theorem 8 in Essay 8) involves a partition of the sequence of partial sums $\{S_n\}$ into two subsequences, one of which is an increasing sequence of lower bounds and the other is a decreasing sequence of upper bounds. The hypotheses for the theorem guarantee that both subsequences converge and to the same value, which implies that the original sequence must converge and to this value. In our generalization, we consider partitions of the original sequence of partial sums into two or more subsequences of a certain type, and prove that if any one of these converges, then the original sequence must also converge, and to the same value. We also try to find one increasing subsequence of lower bounds and one decreasing subsequence of upper bounds, in order to approximate the value of the original sequence with an error estimate.

Theorem 1. (Generalized alternating series test) *Let $\sum_{n=1}^{\infty} a_n$ be a series of mixed signs so that $\lim_{n\to\infty} |a_n| = 0$ and let $S_n = \sum_{i=1}^{n} a_i$. If $\{S_{kn}\}$ converges to S for some $k \in \mathbf{N}$, then $\sum_{n=1}^{\infty} a_n$ also converges to S.*

Proof A proof of this result is based on the fact that $|S - S_{kn}|$ can be made as small as desired for large enough n. For general m, since $kn_o < m < k(n_o + 1)$ for some n_o and since

$$S_m = S_{kn_o} + \sum_{i=kn_o+1}^{m} a_i,$$

it follows that

$$|S - S_m| < |S - S_{kn_o}| + \sum_{i=kn_o+1}^{m} |a_i|.$$

In Exercise 1, you are asked to organize these facts into a careful proof. ∎

Note that this theorem applies to any series, but is not needed for a series of positive terms, where better tests are available. We next provide three examples involving rearrangements of the alternating harmonic series.

Example 1. We begin with the alternating harmonic series

$$\sum_{n=1}^{\infty} a_n = \sum_{n=1}^{\infty} \frac{(-1)^{n-1}}{n} = 1 - \frac{1}{2} + \frac{1}{3} - \frac{1}{4} + \frac{1}{5} - \frac{1}{6} + - \cdots$$

and let $\{A_n\}$ represent its sequence of partial sums and $A = \lim_{n\to\infty} A_n$. By the alternating series test, we know this series converges, having $\{A_{2n}\}$ as an increasing sequence of lower bounds and $\{A_{2n-1}\}$ as a decreasing sequence of upper bounds. The error in using A_{2n} to approximate A is less than

$$|a_{2n+1}| = \frac{1}{2n+1}.$$

We also know this series converges to $\ln 2$ by substituting $x = 1$ into the Maclaurin series

$$\ln(x+1) = x - \frac{x^2}{2} + \frac{x^3}{3} - \frac{x^4}{4} + - \cdots.$$

Example 2. Now consider the rearrangement

$$\sum_{n=1}^{\infty} b_n = 1 + \frac{1}{3} - \frac{1}{2} + \frac{1}{5} + \frac{1}{7} - \frac{1}{4} + \frac{1}{9} + \frac{1}{11} - \frac{1}{6} + + - \cdots$$

and let $\{B_n\}$ where $B_n = b_1 + b_2 + \cdots + b_n$ represent the sequence of partial sums for this series.

Since

$$b_{3i} = -\frac{1}{2i}, \quad b_{3i-1} = \frac{1}{4i-1},$$

and

$$b_{3i-2} = \frac{1}{4i-3},$$

we have

$$b_{3i-2} + b_{3i-1} + b_{3i} = \frac{1}{4i-3} + \frac{1}{4i-1} - \frac{1}{2i} = \frac{8i-3}{(4i-3)(4i-1)(2i)}$$

and so

$$B_{3n} = \sum_{i=1}^{3n} b_i = \sum_{i=1}^{n} (b_{3i-2} + b_{3i-1} + b_{3i})$$

$$= \sum_{i=1}^{n} \frac{8i-3}{(4i-3)(4i-1)(2i)}.$$

We know $\{B_{3n}\}$ converges since the series of positive terms

$$\sum_{i=1}^{\infty} \frac{8i-3}{(4i-3)(4i-1)(2i)}$$

converges by comparison with the series

$$\sum_{i=1}^{\infty} \frac{1}{i^2}.$$

Therefore the series $\sum_{n=1}^{\infty} b_n$ converges by the generalized alternating series test. We next look for ways to approximate the value B to which this series converges. Since

$$\begin{aligned}\sum_{i=1}^{\infty} \frac{8i-3}{(4i-3)(4i-1)(2i)} &= \frac{5}{1\cdot 3\cdot 2} + \frac{13}{5\cdot 7\cdot 4} + \frac{21}{9\cdot 11\cdot 6} + \cdots \\ &= \frac{5}{6} + \frac{13}{140} + \frac{7}{198} + \frac{29}{1560} + \frac{37}{2470} + \cdots,\end{aligned}$$

$\{B_{3n}\} = \{.83333, .92619, .96154, .98013, .99511, \ldots\}$ is an increasing sequence of lower bounds for the value of B.

Because the rearranged series $\sum_{n=1}^{\infty} b_n$ can also be written as

$$\left(1+\frac{1}{3}\right) - \left(\frac{1}{2} - \frac{1}{5} - \frac{1}{7}\right) - \left(\frac{1}{4} - \frac{1}{9} - \frac{1}{11}\right) - \cdots - \left(\frac{1}{2n-2} - \frac{1}{4n-3} - \frac{1}{4n-1}\right)$$

and because

$$\left(\frac{1}{2i-2} - \frac{1}{4i-3} - \frac{1}{4i-1}\right) = \frac{8i-5}{(2i-2)(4i-3)(4i-1)}$$

is positive for all $i > 1$, it follows that $\{B_{3n-1}\}$ is a decreasing sequence of upper bounds for B. Since

$$\begin{aligned}B_{3n-1} &= \frac{4}{3} - \frac{11}{2\cdot 5\cdot 7} - \frac{19}{4\cdot 9\cdot 11} - \frac{27}{6\cdot 13\cdot 15} - \cdots \\ &= \frac{4}{3} - \frac{11}{70} - \frac{19}{396} - \frac{27}{1170} - \frac{35}{2584} - \cdots,\end{aligned}$$

we have $\{B_{3n-1}\} = \{1.33333, 1.17619, 1.12821, 1.10513, 1.09158, \ldots\}$ as the decreasing sequence of upper bounds which approximate B with an error $< |b_{3n}| = 1/2n$.

Using an alternate approach, we have

$$\begin{aligned}B_{3n} = &\left(1+\frac{1}{3}-\frac{1}{2}\right) + \left(\frac{1}{5}+\frac{1}{7}-\frac{1}{4}\right) + \left(\frac{1}{9}+\frac{1}{11}-\frac{1}{6}\right) + \cdots \\ &+ \left(\frac{1}{4n-3} + \frac{1}{4n-1} - \frac{1}{2n}\right)\end{aligned}$$

$$= \left(1 - \frac{1}{2} + \frac{1}{2} + \frac{1}{3} - \frac{1}{4} - \frac{1}{4}\right) + \left(\frac{1}{5} - \frac{1}{6} + \frac{1}{6} + \frac{1}{7} - \frac{1}{8} - \frac{1}{8}\right) + \cdots$$

$$= \left(\left(1 - \frac{1}{2} + \frac{1}{3} - \frac{1}{4}\right) + \left(\frac{1}{2} - \frac{1}{4}\right)\right) + \left(\left(\frac{1}{5} - \frac{1}{6} + \frac{1}{7} - \frac{1}{8}\right) + \left(\frac{1}{6} - \frac{1}{8}\right)\right) + \cdots$$

$$= \left(1 - \frac{1}{2} + \frac{1}{3} - \frac{1}{4} + \frac{1}{5} - \frac{1}{6} + \cdots - \frac{1}{4n}\right) + \left(\frac{1}{2} - \frac{1}{4} + \frac{1}{6} - \frac{1}{8} + \cdots - \frac{1}{4n}\right).$$

As n approaches infinity, the sum in the left parentheses above converges to $\ln 2$ while the sum in the right set of parentheses converges to $\frac{1}{2}\ln 2$. So the entire series converges to $\frac{3}{2}\ln 2 \approx 1.0397$, which is in agreement with the upper and lower bounds we found above.

Example 3. We next consider the rearrangement of $\sum_{n=1}^{\infty} a_n$ where there are three positive terms chosen for each negative one, giving

$$\sum_{n=1}^{\infty} d_n = 1 + \frac{1}{3} + \frac{1}{5} - \frac{1}{2} + \frac{1}{7} + \frac{1}{9} + \frac{1}{11} - \frac{1}{4} + + + - \cdots.$$

Since

$$d_{4i-3} = \frac{1}{6i-5}, \quad d_{4i-2} = \frac{1}{6i-3}, \quad d_{4i-1} = \frac{1}{6i-1}, \quad \text{and} \quad d_{4i} = -\frac{1}{2i},$$

we have

$$d_{4i-3} + d_{4i-2} + d_{4i-1} + d_{4i} = \frac{108i^2 - 92i + 15}{(6i-5)(6i-3)(6i-1)(2i)},$$

and so

$$D_{4n} = \sum_{i=1}^{n} \frac{108i^2 - 92i + 15}{(6i-5)(6i-3)(6i-1)(2i)}.$$

The series $\sum d_n$ converges because

$$\lim_{n\to\infty} D_{4n} = \sum_{n=1}^{\infty} \frac{108n^2 - 92n + 15}{(6n-5)(6n-3)(6n-1)(2n)}$$

and this series converges by comparison with

$$\sum_{n=1}^{\infty} \frac{1}{n^2}.$$

Since

$$D_{4n} = \frac{31}{1\cdot 3\cdot 5\cdot 2} + \frac{263}{7\cdot 9\cdot 11\cdot 4} + \frac{711}{13\cdot 15\cdot 17\cdot 6} + \cdots$$
$$+ \frac{108n^2 - 92n + 15}{(6n-5)(6n-3)(6n-1)(2n)}$$
$$= \frac{31}{30} + \frac{263}{2772} + \frac{711}{19890} + \frac{1375}{73416} + \cdots,$$

it follows that $\{D_{4n}\} = \{1.03333, 1.12821, 1.16395, 1.18268, \ldots\}$ is an increasing sequence of lower bounds for the value of

$$D = \sum_{n=1}^{\infty} d_n.$$

Using a method similar to the one in Example 2, the subsequence $\{D_{4n-1}\}$ is a decreasing sequence of upper bounds because

$$D_{4n-1} = \left(1 + \frac{1}{3} + \frac{1}{5}\right) - \left(\frac{1}{2} - \frac{1}{7} - \frac{1}{9} - \frac{1}{11}\right) - \cdots$$
$$- \left(\frac{1}{2n-2} - \frac{1}{6n-5} - \frac{1}{6n-3} - \frac{1}{6n-1}\right)$$

and because

$$\frac{1}{2n-2} - \frac{1}{6n-5} - \frac{1}{6n-3} - \frac{1}{6n-1} = \frac{108n^2 - 124n + 31}{(2n-2)(6n-5)(6n-3)(6n-1)}$$

is positive for $n > 1$. Since

$$D_{4n-1} = \frac{23}{15} - \frac{215}{1386} - \frac{569}{13260} - \frac{1263}{55062} - \cdots,$$

it follows that a decreasing sequence of lower bounds for the value of D is

$$\{D_{4n-1}\} = \{1.53333, 1.37821, 1.33529, 1.31236, \ldots\}.$$

The next five examples deal with series of mixed signs which are not rearrangements of the alternating harmonic series.

Example 4. Begin with the series of mixed signs defined by

$$\sum_{n=1}^{\infty} f_n = 1 + \frac{1}{2} - \frac{1}{3} + \frac{1}{4} + \frac{1}{5} - \frac{1}{6} + + - \cdots$$

and let $\{F_n\}$ represent the sequence of partial sums. Before we ask about rearrangements of this series, we need to decide whether it converges. To do this, we observe that

$$\begin{aligned} F_{3n} &= \sum_{i=1}^{n} (f_{3i-2} + f_{3i-1} + f_{3i}) \\ &= \sum_{i=1}^{n} \frac{1}{3i-2} + \frac{1}{3i-1} - \frac{1}{3i} \\ &= \sum_{i=1}^{n} \frac{9i^2 - 2}{(3i-2)(3i-1)(3i)}. \end{aligned}$$

Since the series of positive terms

$$\sum_{i=1}^{\infty} \frac{9i^2 - 2}{(3i-2)(3i-1)(3i)}$$

diverges by comparison with the harmonic series

$$\sum_{n=1}^{\infty} \frac{1}{i},$$

it follows that the subsequence $\{F_{3n}\}$ diverges, and so must the series $\sum_{n=1}^{\infty} f_n$. So we do not ask about rearrangements of this series, although it is possible that some of them might converge.

Example 5. Let the series of mixed signs be defined by

$$\sum_{n=1}^{\infty} g_n = 1 + \frac{1}{2} - \frac{1}{3} - \frac{1}{4} + \frac{1}{5} + \frac{1}{6} - \frac{1}{7} - \frac{1}{8} + + - - \cdots$$

and let $\{G_n\}$ represent the sequence of partial sums. Since we can write this series in the form

$$\sum_{n=1}^{\infty}\left(1+\frac{1}{2}\right)-\left(\frac{1}{3}+\frac{1}{4}\right)+\left(\frac{1}{5}+\frac{1}{6}\right)-\left(\frac{1}{7}+\frac{1}{8}\right)+\cdots+(-1)^{n-1}\left(\frac{1}{2n-1}+\frac{1}{2n}\right)+\cdots,$$

we see $\{G_{2n}\}$ is the sequence of partial sums for the convergent alternating series

$$\sum_{n=1}^{\infty} \frac{(-1)^{n-1}4n - 1}{2n(2n-1)},$$

and therefore $\{G_{2n}\}$ converges. This implies that the series $\sum_{n=1}^{\infty} g_n$ converges by Theorem 1. We can approximate the value for this series by the alternating series test, and consider rearrangements of this conditionally convergent series if desired.

Example 6. Notice that the generalized alternating series test can be applied to an alternating series in the usual sense when the decreasing property of $\{|a_n|\}$ does not hold. This usual hypothesis of the alternating series test is only needed to prove that A_n approximates the value for A with an error less than $|a_{n+1}|$. For example, the alternating series test does not apply to the series

$$1 - \tfrac{1}{4} + \tfrac{1}{3} - \tfrac{1}{6} + \tfrac{1}{5} - \tfrac{1}{8} + - \cdots$$

but this series does converge by the generalized alternating series test since the subsequence $\{S_{2n}\}$ where

$$S_{2n} = \sum_{i=2}^{n}\left(\frac{1}{2i-3} - \frac{1}{2i}\right) = \sum_{i=2}^{n} \frac{3}{(2i-3)2i}$$

converges.

Example 7. The alternating series test again fails to apply to the following series because $\{|a_n|\}$ is not monotone decreasing.

$$\frac{1}{3} - \frac{1}{2} + \frac{1}{7} - \frac{1}{4} + \frac{1}{11} - \frac{1}{6} + - \cdots + \frac{1}{4i-1} - \frac{1}{2i} + - \cdots$$

This series diverges because the subsequence $\{S_{2n}\}$ where

$$S_{2n} = \sum_{i=1}^{n} \left(\frac{1}{4i-1} - \frac{1}{2i} \right) = \sum_{i=1}^{n} \frac{1-2i}{(4i-1)(2i)}$$

diverges by comparison with the series

$$\sum_{i=1}^{\infty} \frac{1}{i}.$$

Example 8. Since

$$\sum_{n=1}^{\infty} \frac{1}{n^2}$$

converges, so will any series obtained from this series by making some of the coefficients negative, such as the following.

$$R = \sum_{n=1}^{\infty} r_n = 1 - \frac{1}{2^2} + \frac{1}{3^2} - \frac{1}{4^2} + \frac{1}{5^2} - \frac{1}{6^2} + - \cdots$$

$$S = \sum_{n=1}^{\infty} s_n = 1 + \frac{1}{2^2} - \frac{1}{3^2} + \frac{1}{4^2} + \frac{1}{5^2} - \frac{1}{6^2} + + - \cdots$$

$$T = \sum_{n=1}^{\infty} t_n = 1 - \frac{1}{2^2} - \frac{1}{3^2} + \frac{1}{4^2} - \frac{1}{5^2} - \frac{1}{6^2} + - - \cdots$$

Since each of these three series converges absolutely, all rearrangements of each will also converge, and to the same value. The only task left is to find the values for R, S, and T.

Assuming

$$V = \sum_{n=1}^{\infty} \frac{1}{n^2} = \frac{\pi^2}{6}$$

is known (see Problem 11), we have

$$V - R = 2\left(\frac{1}{2^2} + \frac{1}{4^2} + \cdots \right) = \frac{1}{2}V, \qquad \text{and so} \quad R = \frac{1}{2}V = \frac{\pi^2}{12},$$

$$V - S = 2\left(\frac{1}{3^2} + \frac{1}{6^2} + \cdots \right) = \frac{2}{9}V, \qquad \text{and so} \quad S = \frac{7}{9}V = \frac{7\pi^2}{54}.$$

No such identity applies to the series $\sum_{n=1}^{\infty} t_n$, so its value can only be approximated by the idea that $\{T_{3n-2}\}$ is a decreasing sequence of upper bounds, but no error estimate is available because neither $\{T_{3n}\}$ nor $\{T_{3n-1}\}$ are increasing sequences. This underscores the fact that no one method works for all series, and some series defy all reasonable attempts to approximate its value. If we did not have the result above to show $S = 7\pi^2/54$, we would have similar difficulties in approximating S since $\{S_{3n}\}$ is an increasing sequence of lower bounds, but neither $\{S_{3n-1}\}$ nor $\{S_{3n-2}\}$ provide upper bounds to give an error estimate.

Exercises

1. Provide a careful epsilon proof for Theorem 1.

2. Let an rearrangement of $\sum_{n=1}^{\infty} a_n$ be defined by

$$\sum_{n=1}^{\infty} c_n = 1 - \frac{1}{2} - \frac{1}{4} + \frac{1}{3} - \frac{1}{6} - \frac{1}{8} + \frac{1}{5} - \frac{1}{10} - \frac{1}{12} + - - \cdots,$$

where two negative terms of $\sum_{n=1}^{\infty} a_n$ are chosen for each positive one. Let $\{C_n\}$ represent the sequence of partial sums for this series.

(a) Verify that

$$C_{3n} = \sum_{i=1}^{n} \frac{1}{4i(2i-1)},$$

and explain why this proves $\sum_{n=1}^{\infty} c_n$ converges.

(b) Verify that

$$C_{3n-2} = 1 - \sum_{i=2}^{n} \frac{6i-7}{(2i-3)(2i-1)(4i-4)},$$

and so $\{C_{3n-2}\}$ is a decreasing sequence of upper bounds. Show C_{3n} approximates the value for C with an error less than

$$|c_{3n+1}| = \frac{1}{2n+1}.$$

(c) Explain why the series $\sum_{n=1}^{\infty} c_n$ converges to $\frac{1}{2}\ln 2$. Calculate C_{15} and C_{16} and verify that these lower and upper bounds are consistent with the limit value of $\frac{1}{2}\ln 2$.

3. (a) In Example 5, explain why the series $\sum_{n=1}^{\infty} g_n$ converges to

$$\frac{\pi}{4} + \frac{1}{2}\ln 2.$$

(b) In Example 5, find G_{12} and tell what the error is when using this value to approximate $G = \sum_{n=1}^{\infty} g_n$. Compare this with the value found in part (a).

4. Use the examples above to explain why the following series results are true.

$$\frac{1}{1\cdot 2}+\frac{1}{3\cdot 4}+\frac{1}{5\cdot 6}+\cdots=\ln 2$$
$$\frac{1}{1\cdot 3}+\frac{1}{5\cdot 7}+\frac{1}{9\cdot 11}+\cdots=\frac{\pi}{8}$$
$$\frac{1}{2\cdot 3}+\frac{1}{4\cdot 5}+\frac{1}{6\cdot 7}+\cdots=1-\ln 2$$

Problem 14. Finding Approximations for π

The first use of an infinite series to approximate π was made when Leibniz or James Gregory substituted $x = 1$ into the series expression

$$\arctan x = x - \frac{x^3}{3}+\frac{x^5}{5}-+\cdots=\sum_{n=1}^{\infty}\frac{(-1)^{n-1}x^{2n-1}}{2n-1}$$

in order to obtain the formula

$$\frac{\pi}{4}=1-\frac{1}{3}+\frac{1}{5}-\frac{1}{7}+\cdots.$$

This formula is of limited value because the rate of convergence is so slow (i.e., 500 terms are needed to find an approximation for π with error less than $\frac{1}{1000}$). However within just a few years, John Machin in 1706 obtained the identity

$$\frac{\pi}{4}=4\arctan\frac{1}{5}-\arctan\frac{1}{239},$$

which he then used to approximate π to 100 decimal places. In 1873, William Shanks was able to extend the approximation of π to 707 places by use of the formula discussed in Example 3. Howard Eves provides a very helpful summary of results associated with the number π.[13] The following quote is taken from this summary.

> In 1946, D.F. Ferguson of England discovered errors, starting with the 528th place, in Shanks' value for π, and in January 1947 gave a corrected value to 710 places. In the same month, J.W. Wrench, Jr., of America, published an 808-place value of π, but Ferguson soon found an error in the 723rd place. In January 1948, Ferguson and Wrench jointly published the corrected and checked value of π to 808 places. Wrench used Machin's formula, whereas Ferguson used the formula[14]
>
> $$\frac{\pi}{4}=3\arctan\left(\frac{1}{4}\right)+\arctan\left(\frac{1}{20}\right)+\arctan\left(\frac{1}{1985}\right).$$

The following examples and exercises illustrate how such approximations of π may be obtained. We begin with the notation of a triple (γ,α,β), by which we mean that

[13] See pp. 117–24 in **[11]**.

[14] We obtain this formula in Example 2.

$\arctan\gamma = \arctan\alpha + \arctan\beta$. In Exercise 1, we obtain the result that if γ and α are given, then

$$\beta = \frac{\gamma - \alpha}{1 + \gamma\alpha}.$$

For example, we have $(1, \frac{1}{2}, \frac{1}{3})$ because if $\gamma = 1$ and $\alpha = \frac{1}{2}$, then

$$\beta = \frac{1 - \frac{1}{2}}{1 + \frac{1}{2}} = \frac{1}{3}.$$

We also have $(1, \frac{1}{4}, \frac{3}{5})$ because if $\gamma = 1$ and $\alpha = \frac{1}{4}$, then

$$\beta = \frac{1 - \frac{1}{4}}{1 + \frac{1}{4}} = \frac{3}{5}.$$

We extend this notation to the n-tuple $(\gamma, \alpha_1, \alpha_2, \ldots, \alpha_{n-1})$ which means that

$$\arctan\gamma = \sum_{i=1}^{n-1} \arctan\alpha_i.$$

We can combine the triples $(\gamma, \alpha_1, \alpha_2)$ and $(\alpha_2, \alpha_1, \alpha_3)$ to get the 4-tuple $(\gamma, \alpha_1, \alpha_1, \alpha_3)$. In order to approximate π in a reasonable way from the Maclaurin series for $\arctan x$, we seek an n-tuple where each numerator of α_i is 1 and the denominators are as large as possible.

Example 1. We begin with the triple $(1, \frac{1}{2}, \frac{1}{3})$ obtained above.

Then for $\gamma = \frac{1}{3}$ and $\alpha = \frac{1}{2}$, we have

$$\beta = \frac{\frac{1}{3} - \frac{1}{2}}{1 + \frac{1}{6}} = -\frac{1}{7},$$

and so we have the triple $(\frac{1}{3}, \frac{1}{2}, -\frac{1}{7})$ and by combination, the 4-tuple $(1, \frac{1}{2}, \frac{1}{2}, -\frac{1}{7})$ which means that

$$\frac{\pi}{4} = 2\arctan\frac{1}{2} - \arctan\frac{1}{7}.$$

Observe that $\arctan(-\frac{1}{7}) = -\arctan(\frac{1}{7})$ (see Problem 1 on even and odd functions).

If we try to improve this formula by using $\gamma = \frac{-1}{7}$ and $\alpha = \frac{1}{2}$, we get

$$\beta = \frac{\frac{-1}{7} - \frac{1}{2}}{1 - \frac{1}{14}} = \frac{-9}{13},$$

so we are apparently now moving in the wrong direction.

Example 2. We next repeat the type of calculation done in Example 1, but using the fraction $\frac{1}{4}$ instead of $\frac{1}{2}$. We begin with the triple $(1, \frac{1}{4}, \frac{3}{5})$.

If $\gamma = \frac{3}{5}$ and $\alpha = \frac{1}{4}$, we obtain

$$\beta = \frac{\frac{3}{5} - \frac{1}{4}}{1 + \frac{3}{20}} = \frac{7}{23}$$

which gives the triple $(\frac{3}{5}, \frac{1}{4}, \frac{7}{23})$ and also the 4-tuple $(1, \frac{1}{4}, \frac{1}{4}, \frac{7}{23})$.

Next with $\gamma = \frac{7}{23}$ and $\alpha = \frac{1}{4}$, we get

$$\beta = \frac{\frac{7}{23} - \frac{1}{4}}{1 + \frac{7}{92}} = \frac{5}{99},$$

leading to the triple $(\frac{7}{23}, \frac{1}{4}, \frac{5}{99})$ and also to the 5-tuple $(1, \frac{1}{4}, \frac{1}{4}, \frac{1}{4}, \frac{5}{99})$. The next step gives the 6-tuple

$$\left(1, \frac{1}{4}, \frac{1}{4}, \frac{1}{4}, \frac{1}{4}, \frac{-79}{401}\right),$$

which indicates that matters are getting worse.

At first glance, it seems that none of the above formulas with the $\frac{1}{4}$ fraction give a helpful formula. But if we take the 5-tuple $(1, \frac{1}{4}, \frac{1}{4}, \frac{1}{4}, \frac{5}{99})$ and then set $\gamma = \frac{5}{99}$ and $\alpha = \frac{1}{20}$, we get $\beta = \frac{1}{1985}$, leading to the very helpful 6-tuple

$$\left(1, \frac{1}{4}, \frac{1}{4}, \frac{1}{4}, \frac{1}{20}, \frac{1}{1985}\right)$$

because it translates to the formula

$$\frac{\pi}{4} = 3 \arctan\left(\frac{1}{4}\right) + \arctan\left(\frac{1}{20}\right) + \arctan\left(\frac{1}{1985}\right).$$

We now briefly discuss how to get a triple such as

$$\left(\frac{5}{99}, \frac{1}{20}, \frac{1}{1985}\right).$$

After obtaining the 5-tuple $(1, \frac{1}{4}, \frac{1}{4}, \frac{1}{4}, \frac{5}{99})$, we look for a value $k > 4$ so that when $\gamma = \frac{5}{99}$ and $\alpha = \frac{1}{k}$, then β will have ± 1 in its numerator. Since

$$\beta = \left(\frac{5}{99} - \frac{1}{k}\right) \bigg/ \left(1 + \frac{5}{99k}\right) = \frac{5k - 99}{99k + 5},$$

we see that the numerator $5k - 99$ will equal 1 if $k = 20$.

In general, if an n-tuple is of the form

$$\left(1, \frac{1}{r}, \frac{1}{r}, \ldots, \frac{1}{r}, \frac{a}{b}\right),$$

we look for a value $k > r$ so that if

$$\gamma = \frac{a}{b} \quad \text{and} \quad \alpha = \frac{1}{r},$$

then

$$\beta = \frac{\frac{a}{b} - \frac{1}{k}}{1 + \frac{a}{bk}} = \frac{ak - b}{a + bk}$$

will have a numerator of ± 1 if $ak - b = \pm 1$. This will happen whenever $k = b \pm \frac{1}{a}$ is an integer and is also greater than r. Thus it is clear that a needs to be significantly smaller than b to have any chance of success.

A next logical step is to use values for $\frac{1}{6}$, $\frac{1}{7}$, ... to find a series approximation for π that converges even more rapidly than the ones listed below, which we found by use of the fractions $\frac{1}{2}$, $\frac{1}{3}$, $\frac{1}{4}$ and $\frac{1}{5}$ in Examples 1 and 2 and Exercises 2 and 3.

$$\pi = 8 \arctan\left(\frac{1}{2}\right) - 4 \arctan\left(\frac{1}{7}\right)$$

$$\pi = 8 \arctan\left(\frac{1}{3}\right) + 4 \arctan\left(\frac{1}{7}\right)$$

$$\pi = 12 \arctan\left(\frac{1}{4}\right) + 4 \arctan\left(\frac{1}{20}\right) + 4 \arctan\left(\frac{1}{1985}\right)$$

$$\pi = 16 \arctan\left(\frac{1}{5}\right) - 4 \arctan\left(\frac{1}{239}\right)$$

When we look for a similar formula for π using the fraction $\frac{1}{6}$, we obtain the n-tuples

$$\left(1, \frac{1}{6}, \frac{5}{7}\right), \left(1, \frac{1}{6}, \frac{1}{6}, \frac{23}{47}\right), \left(1, \frac{1}{6}, \frac{1}{6}, \frac{1}{6}, \frac{91}{305}\right) \left(1, \frac{1}{6}, \frac{1}{6}, \frac{1}{6}, \frac{1}{6}, \frac{241}{1921}\right)$$

which do not seem likely to yield a useful formula. A similar solution occurs for the fractions $\frac{1}{7}$, $\frac{1}{8}$ and $\frac{1}{9}$, which suggests we might have already found all the useful formulas. However a formula involving the fraction $\frac{1}{10}$ was unexpectedly found in 1981. This formula is given below as equation (1).

Example 3. We next show how to determine the number of terms needed to obtain an approximation for π with error less than 10^{-101} by using the formula

$$\pi = 12 \arctan\left(\frac{1}{4}\right) + 4 \arctan\left(\frac{1}{20}\right) + 4 \arctan\left(\frac{1}{1985}\right).$$

We ask how many terms of

$$\sum_{n=1}^{\infty} \frac{(-1)^{n+1} 12}{(2n-1)\, 4^{2n-1}}$$

are needed before the error will be less than 10^{-101} (in order to have 100 decimal place accuracy). We need to find n so that

$$\frac{12}{(2n-1)\,4^{2n-1}} < 10^{-101}.$$

To get a ballpark value, we first solve the following inequalities.

$$\frac{1}{4^{2n}} < \frac{1}{10^{101}}$$
$$16^n > 10^{101}$$
$$n > \frac{101 \ln 10}{\ln 16} \approx 83.8.$$

To sharpen the estimate, for $n > 80$, we have

$$\frac{1}{2n-1} < \frac{1}{159}$$

and so

$$\frac{12}{(2n-1)\,4^{2n-1}} = \frac{12 \cdot 4}{(2n-1)\,16^n} < \frac{48}{159} \cdot \frac{1}{16^n} < \frac{1}{10^{101}}$$

which is satisfied if

$$16^n > \frac{48}{159}\,10^{101} \quad \text{or if } n > 83.$$

So our original approximation was very accurate.

Similarly, $n > 38$ must hold for the series associated with the $4\arctan(\frac{1}{20})$ term to give an approximation for π with error less than 10^{-101} and $n > 15$ must hold for the series associated with the $4\arctan(\frac{1}{1985})$ term.

The original series $\pi = \frac{4}{1} - \frac{4}{3} + \frac{4}{5} - \frac{4}{7} + - \cdots$ would have required something of the order of 10^{100} terms to reach the same accuracy achieved by only 84 terms of the series used in Example 3. Even though the approximation obtained by John Machin in 1706 required an even smaller number of terms ($n > 72$), consider the logistical difficulties in actually obtaining this 100 decimal place value by calculating and then adding up these 72 numbers each to 100 decimal places. How would you like to have to calculate just the last term

$$\frac{4}{145 \cdot 5^{145}}$$

to 101 decimal places using only paper and pencil! The power of the computer to dramatically increase the number of digits of accuracy is seen by the fact that only 13 years after Wrench and Ferguson found 808 decimal places of accuracy in 1948, Wrench and Daniel Shanks obtained more than 100,000 decimal places in 1961. And 20 years later in

1981, two Japanese mathematicians found more than two million significant digits using 137 hours of computer time. They used the following formula.

$$\pi = 32\arctan\left(\frac{1}{10}\right) - 4\arctan\left(\frac{1}{239}\right) - 16\arctan\left(\frac{1}{515}\right) \tag{1}$$

The present record is more than 200 billion decimal places of accuracy, which is a long way from the early days of Leibniz and John Machin. With this in mind, one might ask if the efforts of John Machin and others like him were wasted. To get only 700 decimal places of accuracy over a period of 250 years seems as nothing in comparison with the results obtained during the past 50 years. But would these modern successes have occurred without the basis provided for them by Machin and others? And who knows what results will look like 50 years from now!

Exercises

1. Use the trigonometric identity

$$\tan(A+B) = \frac{\tan A + \tan B}{1 - \tan A \tan B}$$

in order to prove that if

$$\arctan\gamma = \arctan\alpha + \arctan\beta,$$

then

$$\beta = \frac{\gamma - \alpha}{1 + \gamma\alpha}.$$

2. Show by successive steps how to obtain the triple $(1, \frac{1}{3}, \frac{1}{2})$, the 4-tuple $(1, \frac{1}{3}, \frac{1}{3}, \frac{1}{7})$, the 5-tuple $(1, \frac{1}{3}, \frac{1}{3}, \frac{1}{3}, \frac{-2}{11})$, and hence the formula that

$$\frac{\pi}{4} = 2\arctan\left(\frac{1}{3}\right) + \arctan\left(\frac{1}{7}\right).$$

3. In this exercise, you will obtain the well-known formula of John Machin. You have learned the process now through a study of Examples 1 and 2 and Exercise 2. Begin with $\gamma = 1$ and $\alpha = \frac{1}{5}$ to get the triple $(1, \frac{1}{5}, \frac{2}{3})$. Repeat this process three more times until you get the 6-tuple

$$\left(1, \frac{1}{5}, \frac{1}{5}, \frac{1}{5}, \frac{1}{5}, -\frac{1}{239}\right)$$

which is equivalent to the formula

$$\frac{\pi}{4} = 4\arctan\left(\frac{1}{5}\right) - \arctan\left(\frac{1}{239}\right).$$

4. Follow the pattern of work just before Example 3 involving the fraction $\frac{1}{6}$ to show that the fraction $\frac{1}{7}$ does not apparently lead to a useful formula either. You may stop after obtaining the following n-tuples.

$$\left(1, \frac{1}{7}, \frac{3}{4}\right) \quad \left(1, \frac{1}{7}, \frac{1}{7}, \frac{17}{31}\right) \quad \left(1, \frac{1}{7}, \frac{1}{7}, \frac{1}{7}, \frac{44}{117}\right) \quad \text{and} \quad \left(1, \frac{1}{7}, \frac{1}{7}, \frac{1}{7}, \frac{1}{7}, \frac{119}{863}\right)$$

5. Show how to obtain the formula

$$\frac{\pi}{4} = \arctan\left(\frac{1}{2}\right) + \arctan\left(\frac{1}{5}\right) + \arctan\left(\frac{1}{8}\right),$$

which was first found in 1844. Begin with the formula

$$\frac{\pi}{4} = \arctan\left(\frac{1}{2}\right) + \arctan\left(\frac{1}{3}\right)$$

obtained earlier in this problem.

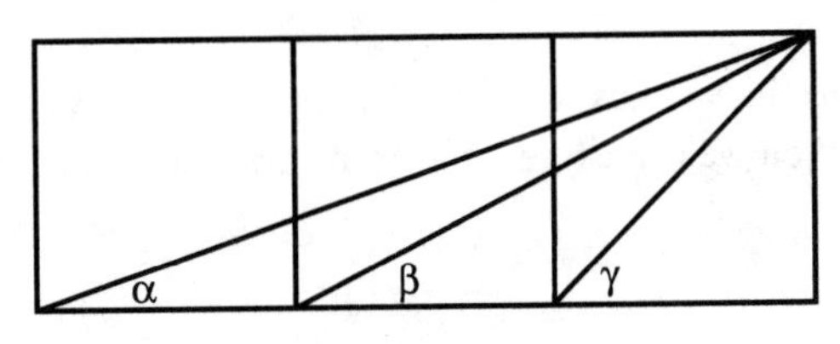

FIGURE 2

6. We established all the formulas for $\pi/4$ in this problem by the use of trigonometry. For contrast, establish the formula

$$\frac{\pi}{4} = \arctan\left(\frac{1}{2}\right) + \arctan\left(\frac{1}{3}\right)$$

by the use of a geometrical argument. Use Figure 2 consisting of three adjacent squares to show that $\gamma = \alpha + \beta$.[15]

7. The formula

$$\frac{\pi}{4} = 4\arctan\left(\frac{1}{5}\right) - \arctan\left(\frac{1}{70}\right) + \arctan\left(\frac{1}{99}\right)$$

was obtained in 1841. Show how to obtain this formula, beginning with Machin's formula

$$\frac{\pi}{4} = 4\arctan\left(\frac{1}{5}\right) - \arctan\left(\frac{1}{239}\right).$$

8. By using the formula Machin developed in 1706 to approximate π to 100 decimal places, show that 73 terms need to be summed so that the series expression of

[15] A proof may be found in *Mathematical Gazette*, 57 (1973) p. 334.

$$16 \arctan\left(\frac{1}{5}\right) = \sum_{n=1}^{\infty} \frac{(-1)^{n-1}16}{(2n-1)5^{2n-1}}$$

will approximate π with an error less than 10^{-101}. Compare your answer with the result obtained in Example 3. How many terms are needed if the approximation uses the series expression for

$$4 \arctan\left(\frac{1}{239}\right)?$$

9. In 1873, William Shanks found an approximation for π accurate to 707 decimal places using the formula that Machin obtained to find 100 decimal places for π in 1706. Show that Shanks would need about 526 terms to obtain this degree of accuracy for the series expresssion of $16 \arctan(\frac{1}{5})$ given in Exercise 8.

Problem 15. A Function with Derivatives of All Orders, but No Maclaurin Series

Based on our experience in calculus, we have confidence that if we can find the values for $f^{(n)}(0)$ for all $n \in \mathbf{N}$, then we can express the Maclaurin series for $f(x)$ by

$$\sum_{n=0}^{\infty} \frac{f^{(n)}(0)}{n!} x^n.$$

Sometimes, we are able to find a Maclaurin series for a function $f(x)$ even though we cannot directly calculate all the values for $f^{(n)}(0)$. An example of this is $f(x) = \arctan x$ (see Problem 3). The function $g(x) = e^{-1/x^2}$ illustrates a different situation, where we can find values for each $g^{(n)}(0)$, but do not get a Maclaurin series expression for $g(x)$ from it. The goal of this problem is to show that $g^{(n)}(0) = 0$ for all n. So even though $g(x) = e^{-1/x^2}$ is a well-defined continuous function (if we define $g(0) = \lim_{x\to 0} e^{-1/x^2} = 0$), $g(x)$ is not represented by

$$\sum_{n=0}^{\infty} \frac{g^{(n)}(0)}{n!} x^n.$$

Exercises

1. Calculate $g'(0)$ by each of the following two methods.

   (a) $g'(0) = \lim_{x\to 0} \dfrac{g(x) - g(0)}{x - 0}$ (using the definition of the derivative)

   (b) $g'(0) = \lim_{x\to 0} g'(x)$ (using the definition of continuity)

   In either case, you need L'Hospital's rule. Show first that the usual application of this rule fails to give a conclusion. Then use a substitution such as $t = 1/x^2$ to successfully show that the limit is zero.

2. Now show that $g''(0) = 0$ by an argument similar to either one in Exercise 1.

3. Based on your work in Exercises 1 and 2, explain why $g^{(n)}(0) = 0$ for all $n \geq 3$. You do not need to carry out all the calculations.

4. Explain how the usual statement given above about the Maclaurin series formula for $f(x)$ needs to be clarified in light of this example.

Problem 16. Evaluating $\int_0^1 x^k\,dx$ by Fermat's Method

Fermat was interested in finding the same area formulas as was Cavalieri (see Problem 4), namely the area under $f(x) = x^k$ and above the interval $[0, 1]$. But instead of using a subdivision of the interval into n equal subintervals as Cavalieri did, Fermat introduced a different partition. His choice led to a geometric sequence which avoided the complicated sums that Cavalieri obtained. Fermat chose a real number p less than but close to 1, and subdivided the interval $[0, 1]$ by the partition

$$P = \{0, p^n, p^{n-1}, \ldots, p^2, p, 1\}.$$

Using the left endpoints to determine the height of the approximating rectangles, Fermat determined the lower bound sum (by summing from right to left) to be

$$L(f, P) = (1 - p)f(p) + (p - p^2)f(p^2) + \cdots + (p^{n-1} - p^n)\,f(p^n) + (p^n - 0)\,f(0).$$

Example 1. For $f(x) = x^2$, instead of Cavalieri's sum[16]

$$U(f, P_n) = \sum_{i=1}^{\infty} \left(\frac{i}{n}\right)^2 \cdot \frac{1}{n},$$

we now have

$$\begin{aligned}
L(f, P) &= (1 - p)p^2 + (p - p^2)p^4 + (p^2 - p^3)p^6 + \cdots + (p^{n-1} - p^n)p^{2n} + p^n \cdot 0 \\
&= (1 - p)p^2\left(1 + p^3 + p^6 + \cdots + p^{n-1} \cdot p^{2n-2}\right) \\
&= (1 - p)p^2\left(1 + p^3 + p^6 + \cdots + p^{3(n-1)}\right) \\
&= (1 - p)p^2 \frac{1 - p^{3n}}{1 - p^3} = \frac{p^2(1 - p^{3n})}{1 + p + p^2}
\end{aligned}$$

and so

$$\lim_{p \to 1}\left\{\lim_{n \to \infty} L(f, P)\right\} = \frac{1}{3}.$$

[16] See Example 1 in Problem 4.

Exercises

1. (a) Find the expression for $L(f, P)$ when $f(x) = x^4$. Then use the geometric sequence formula to reduce your answer to the form

$$(1-p)p^4 \frac{1-p^{5n}}{1-p^5}.$$

(b) Simplify your answer in part (a) by first assuming that n approaches infinity. Then use cancellation to show that as p approaches 1, the approximating sum $L(f, P)$ approaches $\frac{1}{5}$. Compare the amount of work needed here with Cavalieri's method in Exercise 1 from Problem 4.

2. (a) Show that for the general case of $f(x) = x^k$, the above formula becomes

$$L(f, P) = (1-p)\, p^k \frac{1-(p^{k+1})^n}{1-p^{k+1}}.$$

(b) Take the limit as $n \to \infty$ to obtain

$$\frac{(1-p)p^k}{1-p^{k+1}}.$$

Finally, take the limit as $p \to 1$ to obtain the value of $1/(k+1)$ for the desired area. Hint: Consider the factorization of $1 - p^{k+1}$.

3. Clearly, Fermat's method is superior to that of Cavalieri because it can be applied directly to any positive integer k without the need for first obtaining each separate summation formula for all values $n < k$. Does Fermat's method apply to negative integer, or rational, values for k? Try $k = \frac{1}{2}$.

Problem 17. Two Proofs that $\int_0^1 (x \ln x)^n \, dx = (-1)^n n!/(n+1)^{(n+1)}$

This result was known to John Bernoulli during the 1680s because he used it as part of his remarkable evaluation of the integral $\int_0^1 x^x \, dx$ (see Problem 18). He most likely found this result by use of integration by parts, as discussed in Method 1 below. A more elegant approach uses the gamma function (see Problem 29), as presented in Method 2.

Method 1. We might think this result can be verified by use of mathematical induction, but this is not the case. The recursive formula we need requires beginning with different exponents for the x and the $\ln x$ terms. If you doubt this, try Exercise 1. Exercises 2 to 4 outline the work necessary to obtain the desired formula.

Method 2. We use the gamma function, defined by

$$\Gamma(x) = \int_0^\infty t^{x-1} e^{-t} \, dt,$$

along with the well-known result that $\Gamma(n) = (n-1)!$ for positive integers n. The main steps for this method of evaluating

$$\int_0^1 (x \ln x)^n\,dx$$

are presented in Exercise 5.

Exercises

1. Use integration by parts to develop the formula

$$\int x^n (\ln x)^n\,dx = \frac{x^{n+1}(\ln x)^n}{n+1} - \frac{n}{n+1}\int x^n (\ln x)^{n-1}\,dx.$$

Convince yourself that it is not possible to use this to obtain a unique expression for

$$\int x^n (\ln x)^{n-1}\,dx.$$

See Exercise 2 for additional insight.

2. Use integration by parts to develop the recursion formula

$$\int x^r (\ln x)^t\,dx = \frac{x^{r+1}(\ln x)^t}{r+1} - \frac{t}{r+1}\int x^r (\ln x)^{t-1}\,dx.$$

Notice that this formula permits us to also obtain

$$\int x^r (\ln x)^{t-1}\,dx = \frac{x^{r+1}(\ln x)^{t-1}}{r+1} - \frac{t-1}{r+1}\int x^r (\ln x)^{t-2}\,dx.$$

3. Use L'Hospital's rule to verify that $\lim_{x\to 0^+} x^m (\ln x)^n = 0$ for all $m, n > 0$.

4. Use the results in Exercises 2 and 3 to establish

$$\int_0^1 x^r (\ln x)^t\,dx = \left(-\frac{t}{r+1}\right)\left(-\frac{t-1}{r+1}\right)\left(-\frac{t-2}{r+1}\right)\cdots\left(-\frac{1}{r+1}\right)\int_0^1 x^r\,dx$$

$$= \frac{(-1)^t t!}{(r+1)^t}\,\frac{1}{r+1}.$$

If $r = t = n$, we have the desired formula.

5. Verify the correctness of each of the following steps.

$$\int_0^1 (x \ln x)^n\,dx = \int_\infty^0 (-te^{-t})^n(-e^{-t}\,dt) \quad \text{using the substitution } x = e^{-t}$$

$$= (-1)^n \int_0^\infty e^{-(n+1)t} t^n\,dt$$

$$= (-1)^n \int_0^\infty e^{-s} \frac{s^n}{(n+1)^n} \cdot \frac{ds}{n+1} \qquad \text{using } s = (n+1)t$$

$$= \frac{(-1)^n}{(n+1)^{n+1}} \int_0^\infty s^n e^{-s}\, ds$$

$$= \frac{(-1)^n}{(n+1)^{n+1}} \Gamma(n+1)$$

$$= \frac{(-1)^n}{(n+1)^{n+1}} n! \quad \text{which is the desired formula.}$$

Problem 18. Evaluating $\int_0^1 x^x\,dx$ by Bernoulli's Method

In Problems 4 and 16, we examined the methods used by Cavalieri in the 1630s and by Fermat in the 1640s to evaluate the quantity we now refer to as $\int_0^1 x^k\,dx$. These complicated methods were rendered unnecessary when Newton and Leibniz discovered that antiderivatives could be used to solve such problems quickly and easily. This method does not help with the more complicated integral $\int_0^1 x^x dx$ because there is no simple antiderivative for this integrand function. However, in the 1680s, John Bernoulli found an ingenious way to approximate this integral to any degree of accuracy. His solution is broken down into the following exercises and observations. Convince yourself of the correctness of each one.

Exercises

1. Recall that the Maclaurin series expansion for e^x is

$$\sum_{n=0}^{\infty} \frac{x^n}{n!}.$$

2. Observe that $x^x = e^{x \ln x}$ by a logarithm identity.

3. Verify that the series

$$\sum_{n=1}^{\infty} (x \ln x)^n$$

converges uniformly on the interval [0, 1] by use of the Weierstrass M-test (see Problem 21).

4. Verify that

$$\int_0^1 (x \ln x)^n\, dx = \frac{(-1)^n n!}{(n+1)^{n+1}}$$

(use Problem 17).

5. Use the alternating series test to find a value for the series

$$\sum_{n=0}^{\infty} \frac{(-1)^n}{(n+1)^{n+1}}$$

with an error less than 10^{-6}.

6. Explain how the above results are used to verify the following six steps.

$$\begin{aligned}
\int_0^1 x^x \, dx &= \int_0^1 e^{x \ln x} \, dx && (1) \\
&= \int_0^1 \sum_{n=0}^{\infty} \frac{(x \ln x)^n}{n!} \, dx && (2) \\
&= \sum_{n=0}^{\infty} \int_0^1 \frac{(x \ln x)^n}{n!} \, dx && (3) \\
&= \sum_{n=0}^{\infty} \frac{(-1)^n}{(n+1)^{n+1}} && (4) \\
&= 1 - \frac{1}{2^2} + \frac{1}{3^3} - \frac{1}{4^4} + - \cdots && (5) \\
&\approx .7834305 && (6)
\end{aligned}$$

7. By a similar argument, prove the related formula

$$\int_0^1 \frac{1}{x^x} \, dx = \sum_{n=1}^{\infty} \frac{1}{n^n}.$$

Problem 19. Some Interesting Sequences of Functions

When illustrating sequences of functions, most textbook examples are such that the limit function $f(x)$ is fairly trivial, being essentially constant. The following exercises provide two examples of a sequence of functions where the limit function is more interesting.

Exercises

1. **(a)** Define the sequence $f_n(x) = n(\sqrt[n]{x} - 1)$. Sketch the graphs for f_1, f_2, f_3, and f_4.

   (b) $f(x) = \lim_{n\to\infty} f_n(x)$ is a well-known function. Use your graphs in part (a) to make a guess of what $f(x)$ might be. Then provide a proof for your conjecture using the following two methods.

   i. One method uses the sequence of integrals

$$\int_1^x f_n'(t) \, dt$$

and the result that

$$\lim_{n\to\infty}\int_1^x f_n'(t)\,dt = \int_1^x \lim_{n\to\infty} f_n'(t)\,dt.$$

ii. A second method uses L'Hospital's rule to find $\lim_{n\to\infty} f_n(x)$.

2. A sequence of functions $\{f_n(x)\}$ is defined recursively by the formula below.

$$f_{n+1}(x) = \frac{1}{x}\left(f_n(x) + \frac{1}{f_n(x)}\right) \quad \text{where } f_0(x) = x$$

(a) Show that

$$f_2(x) = \frac{1}{x} + \frac{1}{x^3} + \frac{x}{x^2+1}.$$

(b) Assuming that the limit exists and is not zero, find the formula for

$$f(x) = \lim_{n\to\infty} f_n(x).$$

Problem 20. Comparing Uniform Continuity and Uniform Convergence

Uniform continuity and uniform convergence are important concepts in analysis, providing the means to prove a number of useful theorems. The purpose of this problem is to study the similarities and differences of these two parallel concepts. In each empty box below, you should write in the statement that is analogous to the one next to it. Be as complete as possible with your responses.

Uniform Continuity	Uniform Convergence
	This concept is applied to a sequence of functions $\{f_n(x)\}$ that are defined on a set S.
By definition, this concept requires that for every $\epsilon > 0$, there is a uniform value of δ with the property that $\lvert f(x) - f(u)\rvert < \epsilon$ for all $x, u \in S$ when $\lvert x - u\rvert < \delta$.	
	The way(s) to verify there is uniform convergence is to find a sequence $\{t_n\}$ so that $\lim_{n\to\infty} t_n = 0$ and $\lvert f_n(x) - f(x)\rvert \le t_n$ for all $x \in S$.
The way(s) to show that uniform continuity does not hold is to find sequences $\{x_n\}$ and $\{u_n\}$ in S so that $\lim_{n\to\infty} \lvert x_n - u_n\rvert = 0$, but $\lvert f(x_n) - f(u_n)\rvert \ge \epsilon_o$.	
The main theorem(s) obtained from the use of this concept is that if $f(x)$ is uniformly continuous on $[a, b]$, then $f(x)$ is integrable on $[a, b]$.	

Problem 21. Questions for Sequences of Functions and for Series of Functions

A large part of real analysis is concerned with sequences of functions and series of functions. It is helpful to enumerate and distinquish between the questions needed to investigate a sequence of functions compared with those needed to study a series of functions. The latter include the special cases of power series and Fourier series.

Topic 1. Sequences of Functions We begin with a sequence of functions $\{f_n(x)\}$ and seek answers to the following three questions.

Question 1. What is the limit function

$$f(x) = \lim_{n\to\infty} f_n(x)?$$

The limit function is found by a variety of methods, including algebraic identities and the use of L'Hospital's rule. It is essential to find an expression for $f(x)$ before dealing with Question 3.

Question 2. What is the domain D of pointwise convergence for $f(x)$?

The answer for this question is usually found in conjunction with the work for Question 1.

Question 3. For what subsets of D is the convergence of $\{f_n(x)\}$ to $f(x)$ uniform?

This question is usually answered by the definition of convergence, namely deciding for which sets $S \subset D$ it is true that $|f_n(x) - f(x)| < \epsilon$ for all $n \geq N$ and for all $x \in S$. A helpful shortcut is to find a sequence of constants $\{t_n\}$ so that

$$\lim_{n\to\infty} t_n = 0 \quad \text{and} \quad |f_n(x) - f(x)| \leq t_n \quad \text{for all } x \in S.$$

The standard theorems involving uniform convergence guarantee such results as the following. If all the $f_n(x)$ functions are continuous on S, then

1. $f(x)$ is continuous on S and
2. $\displaystyle\lim_{n\to\infty} \int_a^b f_n(x)\,dx = \int_a^b f(x)\,dx$ for any interval $[a, b] \subset S$.

Example 1. Let the given sequence be

$$f_n(x) = \frac{x^n}{1 + x^n}.$$

The only undefined values are at $x = -1$ for odd values of n. It is straightforward to see that

$$f(x) = \lim_{n\to\infty} f_n(x) = 1 \qquad \text{if } |x| > 1$$

and

$$f(x) = \lim_{n\to\infty} f_n(x) = 0 \qquad \text{if } |x| < 1.$$

Also $f(1) = \frac{1}{2}$ and $f(-1)$ is not defined. So $f(x)$ is defined for all reals except $x = -1$ and has a jump discontinuity at $x = 1$. The convergence is uniform on sets of the form

$$(-\infty, -1+\epsilon], \quad [-1+\epsilon, 1-\epsilon], \quad \text{and} \quad [1+\epsilon, \infty)$$

for small values of $\epsilon > 0$. In order to prove $\{f_n(x)\}$ converges uniformly to 1 on the set $[t, \infty)$ for $t > 1$, we observe that

$$|f_n(x) - f(x)| = \left| \frac{x^n}{1+x^n} - 1 \right| = \frac{1}{1+x^n} \leq \frac{1}{1+t^n}$$

for all

$$x \geq t \quad \text{and} \quad \lim_{n\to\infty} \frac{1}{1+t^n} = 0.$$

Topic 2. Series of Functions We consider a different set of questions for the series of functions $\sum_{n=1}^{\infty} u_n(x)$.

Question 1. For which values of x does the series converge?

This is typically answered using the ratio test, or the root test in some cases. In distinction to the case above for sequences of functions, we are usually unable to find a simple expression for

$$u(x) = \sum_{n=1}^{\infty} u_n(x),$$

unless geometric series are involved.

Question 2. For which set S is the convergence uniform?

The usual method for answering this question is the Weierstrass M-test. We base our guess for the set S where the convergence is uniform on prior knowledge of the set D of pointwise convergence found in Question 1.

Because the results under Topic 1 can be applied to the sequence of partial sums $\{f_n(x)\}$ where

$$f_n(x) = \sum_{i-1}^{n} u_i(x),$$

we have such new results available as

$$\int_a^b \sum_{n=1}^{\infty} u_n(x)\,dx = \sum_{n=1}^{\infty} \int_a^b u_n(x)\,dx$$

if the series

$$\sum_{n=1}^{\infty} u_n(x)$$

converges uniformly on $[a, b]$.

Example 2. In Problem 18, we needed the result that the series of functions

$$\sum_{n=1}^{\infty} u_n(x) = \sum_{n=1}^{\infty} (x \ln x)^n$$

converges uniformly on the interval $[0, 1]$. To accomplish this, we show the function $y = x \ln x$ has a relative minimum at

$$\left(\frac{1}{e}, -\frac{1}{e}\right)$$

and that

$$|x \ln x| \leq \frac{1}{e} \quad \text{for } x \in [0, 1].$$

Therefore the series

$$\sum_{n=1}^{\infty} (x \ln x)^n$$

converges uniformly on $[0, 1]$ by the Weierstrass M-test with $M_n = (1/e)^n$.

We next observe some additional results when dealing with the special cases of a power series or a Fourier series.

Topic 3. Power series When we consider a general power series

$$\sum_{n=1}^{\infty} a_n x^n,$$

we ask the same two questions as in Topic 2. The answer to the first question is that the set of pointwise convergence is an interval centered about $x = 0$. (For the more general power series of the form

$$\sum_{n=1}^{\infty} a_n (x - c)^n,$$

the interval of convergence will be centered about $x = c$). The interval may be finite or infinite, and it may or may not include the endpoints. The set of uniform convergence will also be an interval. For example, if the interval of pointwise convergence is $(-a, a]$, then the sets of uniform convergence will be intervals of the form $[-a + \epsilon, a]$ for small values of $\epsilon > 0$.

Assume the function $f(x)$ is represented by the Maclaurin series

$$\sum_{n=0}^{\infty} \frac{f^{(n)}(0)}{n!} x^n$$

on the interval I. A new third question arises in this case.

Question 3. Does the series converge to the same value as given by the function $f(x)$ that determined the series?

Problem 15 shows the answer to this question may be no. In order to answer this question in the affirmative, we need to show that the remainder term from Taylor's theorem

$$\left(R_n(x) = \frac{f^{(n+1)}(c)}{(n+1)!} x^{n+1} \quad \text{for some value } c \text{ between 0 and } x \right)$$

approaches zero for all x in the interval of convergence. The details for this vary with each function.

Example 3. The integral $\int_0^1 e^{-x^2} dx$ cannot be evaluated by the fundamental theorem of calculus since there is no simple antiderivative, but because the Maclaurin series

$$e^x = \sum_{n=0}^{\infty} \frac{x^n}{n!}$$

converges uniformly for all reals, we have the following equality.

$$\begin{aligned}
\int_0^1 e^{-x^2} dx &= \int_0^1 \sum_{n=0}^{\infty} \frac{(-x^2)^n}{n!} \\
&= \sum_{n=0}^{\infty} \int_0^1 \frac{(-1)^n x^{2n}}{n!} dx \\
&= \sum_{n=0}^{\infty} \frac{(-1)^n x^{2n+1}}{n!(2n+1)} \Bigg|_0^1 \\
&= \sum_{n=0}^{\infty} \frac{(-1)^n}{n!(2n+1)} \\
&= 1 - \frac{1}{3} + \frac{1}{10} - \frac{1}{42} + \frac{1}{216} - \frac{1}{1320} + \frac{1}{9360} - + \cdots
\end{aligned}$$

Topic 4. Fourier series We can express the Fourier series expansion for $f(x)$ in the form

$$\sum_{n=1}^{\infty} A_n \sin nx + \sum_{n=0}^{\infty} B_n \cos nx$$

by use of standard formulas for A_n and B_n as found in Problem 31. The convergence question for Fourier series is more difficult than the one for power series because the sine and cosine functions are more complicated than polynomials. Also the function $f(x)$ may have many points of discontinuity, which is never the case for power series representations. In addition to verifying that a trigonometric series converges for given values of x, we need to verify it converges to the same value as given by $f(x)$. If $f(x)$ is discontinuous at some x_o, further adjustments are needed.

Exercises

1. In Example 1, verify that the sequence

$$\left\{\frac{x^n}{1+x^n}\right\}$$

converges uniformly to 0 on sets of the form $[0, b]$ for $0 < b < 1$.

2. Answer the three questions under Topic 1 for the sequence $\{u_n(x)\}$ where

$$u_n(x) = \frac{nx}{1+nx}.$$

You should be able to prove the convergence is uniform on sets of the form $[t, \infty)$ for $t > 0$.

3. **(a)** Verify that the series

$$\sum_{n=1}^{\infty} ne^{-nx}$$

converges uniformly on sets of the form $[t, \infty)$ for $t > 0$.

(b) Use the result in part (a) and a geometric series result to show

$$\int_1^2 \sum_{n=1}^{\infty} ne^{-nx}\, dx = \frac{e}{e^2-1}.$$

4.

$$\sum_{n=1}^{\infty} \frac{\cos nx}{n^2}$$

is a Fourier series representation of some even function. Give all necessary reasons to verify

$$\int_0^{\pi/2} \sum_{n=1}^{\infty} \frac{\cos nx}{n^2}\, dx = \sum_{n=1}^{\infty} \frac{(-1)^{n+1}}{(2n-1)^3}.$$

You need to show the convergence is uniform on $[0, \frac{\pi}{2}]$ and then interchange the order of summation and integration.

Section 3. Enrichment Problems

Problem 22. Generalizing the Fibonacci Sequence

The Fibonacci sequence $\{F_n\}$ is defined by the equation $F_n = F_{n-1} + F_{n-2}$ for $n > 2$ where $F_1 = 1$ and $F_2 = 1$. It has been studied during the past 800 years from the time when Leonardo of Pisa included it in *The Book of Squares* which he published in 1202. In analysis, it is not usually considered to be an interesting sequence because it diverges to infinity. However, some variations of this sequence are of interest to us.

Example 1. The sequence $\{x_n\}$ defined by $x_n = \frac{1}{2}(x_{n-1} + x_{n-2})$ for $n > 2$ with $x_1 = 1$, $x_2 = 2$ has a monotone increasing subsequence $\{x_{2n-1}\}$ and a monotone decreasing subsequence $\{x_{2n}\}$. Both subsequences, and hence the original sequence, can be shown to converge to $\frac{5}{3}$.

Solution. Using the recursion formula, we calculate the first several terms.

$$x_3 = \frac{1}{2}(2+1) = \frac{3}{2}$$

$$x_4 = \frac{1}{2}\left(\frac{3}{2} + 2\right) = \frac{7}{4} = 1 + \frac{1}{2} + \frac{1}{4}$$

$$x_5 = \frac{1}{2}\left(\frac{7}{4} + \frac{3}{2}\right) = \frac{13}{8} = 1 + \frac{1}{2} + \frac{1}{4} - \frac{1}{8}$$

$$x_6 = \frac{1}{2}\left(\frac{13}{8} + \frac{7}{4}\right) = \frac{27}{16} = 1 + \frac{1}{2} + \frac{1}{4} - \frac{1}{8} + \frac{1}{16}$$

$$x_7 = \frac{1}{2}\left(\frac{27}{16} + \frac{13}{8}\right) = \frac{53}{32}$$

$$x_8 = \frac{1}{2}\left(\frac{53}{32} + \frac{27}{16}\right) = \frac{107}{64}$$

So the subsequence

$$\{x_{2n}\} = \left\{2, \frac{7}{4}, \frac{27}{16}, \frac{107}{64}, \dots\right\}$$

is decreasing and bounded below, and therefore convergent. Similarly, the subsequence

$$\{x_{2n-1}\} = \left\{1, \frac{3}{2}, \frac{13}{8}, \frac{53}{32}, \dots\right\}$$

is increasing and bounded above, and hence convergent. We further observe that

$$\begin{aligned}
x_{2n} &= 2 - \frac{1}{4} - \frac{1}{16} - \frac{1}{64} - \cdots - \frac{1}{4^{n-1}} \\
&= 2 - \frac{1}{4}\left(1 + \frac{1}{4} + \frac{1}{16} + \cdots + \frac{1}{4^{n-2}}\right) \\
&= 2 - \frac{1}{4} \cdot \frac{1 - \frac{1}{4^{n-1}}}{1 - \frac{1}{4}}
\end{aligned}$$

and so

$$\lim_{n\to\infty} x_{2n} = 2 - \frac{1}{4} \cdot \frac{1}{1-\frac{1}{4}} = \frac{5}{3}.$$

In a similar manner,

$$\begin{aligned} x_{2n-1} &= 1 + \frac{1}{2} + \frac{1}{8} + \frac{1}{32} + \cdots + \frac{1}{2^{2n-3}} \\ &= 1 + \frac{1}{2}\left(1 + \frac{1}{4} + \frac{1}{16} + \cdots + \frac{1}{4^{n-2}}\right) \\ &= 1 + \frac{1}{2} \cdot \frac{1 - \frac{1}{4^{n-1}}}{1 - \frac{1}{4}} \end{aligned}$$

and so

$$\lim_{n\to\infty} x_{2n-1} = 1 + \frac{1}{2} \cdot \frac{1}{1-\frac{1}{4}} = \frac{5}{3}.$$

Therefore the sequence $\{x_n\}$ must converge to $\frac{5}{3}$. ■

Exercises

1. Use the pattern suggested by the expressions for x_4, x_5, and x_6 in Example 1 to find $\lim_{n\to\infty} x_n$ in another way.

2. Let the sequence $\{x_n\}$ be defined recursively by $x_n = \frac{1}{3}x_{n-1} + \frac{2}{3}x_{n-2}$ for $n > 2$, where $x_1 = 1$ and $x_2 = 2$. Use a method involving geometric series as illustrated in Example 1 to show that $\{x_n\}$ converges to $\frac{8}{5}$.

3. Let the sequence $\{x_n\}$ be defined by $x_n = \frac{1}{2}(x_{n-1} + x_{n-2})$ for $n > 2$ where $x_1 = a$ and $x_2 = b$. Show that $\{x_n\}$ converges to $(2b + a)/3$.

4. The sequence $\{x_n\}$ defined by $x_n = a\,x_{n-1} + b\,x_{n-2}$ for $n > 2$ will converge if $a+b = 1$, assuming that $0 < a < 1$ and $0 < b < 1$. Prove that the sequence $\{x_n\}$ converges to
$$\frac{bx_1 + x_2}{1+b}$$

5. Generalize the problems in Exercises 2 and 3 to the more difficult one of evaluating the sequence $\{y_n\}$ defined by $y_n = \frac{1}{3}(y_{n-1} + y_{n-2} + y_{n-3})$ for $n > 3$, if $y_1 = 1$, $y_2 = 2$, and $y_3 = 3$.

Problem 23. Euler's Proof that *e* Is Irrational

The Greek view of mathematics was significantly altered around 500 B.C. with the discovery by a follower of Pythagoras that $\sqrt{2}$ was not a rational number. More than two thousand years passed before another proof of irrationality was successfully completed. In

1763, Euler proved during the period of 25 years he spent at the Berlin Academy that e was irrational.[17] An associate of Euler at the Berlin Academy, Johann Lambert (1728–77) proved in 1770 that π was irrational. Euler's constant γ is another important number that dates from this time period, but we still do not know whether it is rational or irrational. A more recent example is the proof in 1978 by Roger Apéry (1916–1994) that the value for the series

$$\sum_{n=1}^{\infty} \frac{1}{n^3}$$

is an irrational number.[18]

More than 100 years after these proofs by Euler and Lambert, e was proved to be transcendental by Charles Hermite (1822–1901) in 1873, and π was proved transcendental by Ferdinand Lindemann (1852–1939) in 1882. A number is *transcendental* if it is not algebraic, which means it is not a zero of any polynomial with integer coefficients.

Euler used a Maclaurin series as the key element in his proof, a familiar technique of his. We saw in Problem 11 how he used the Maclaurin series for $\sin x$ in his proof that

$$\sum_{n=1}^{\infty} \frac{1}{n^2} = \frac{\pi^2}{6}.$$

Supply a reason for each step of Euler's proof given in Theorem 1.

Theorem 1. *e is irrational.*

Proof As is done for the proof that $\sqrt{2}$ is irrational, we begin by denying the desired conclusion and assuming that $e = a/b$ where a and b are positive integers. Recall the Maclaurin series expansion that

$$e^x = 1 + x + \frac{x^2}{2!} + \frac{x^3}{3!} + \cdots + \frac{x^n}{n!} + \cdots.$$

If $x = -1$, we obtain

$$\frac{1}{e} = 1 - 1 + \frac{1}{2!} - \frac{1}{3!} + \cdots \frac{(-1)^n}{n!} + \cdots.$$

By use of alternating series results, we have

$$0 < \frac{b}{a} - \frac{1}{2!} + \frac{1}{3!} - \cdots + \frac{1}{(n-1)!} < \frac{1}{n!} \quad \text{for } n \text{ even.}$$

[17]For an extended discussion about this number, see Eli Maor, *e: The Story of a Number*, Princeton University Press, Princeton, New Jersey, 1994.

[18]Alfred van der Poorten, "A Proof that Euler Missed," *The Mathematical Intelligencer*, Vol 1, No. 4, 1978, pp. 195–203.

Choose n large enough so that $a|n!$ and then multiply the inequality above by $n!$ to get

$$0 < \frac{b\,n!}{a} - \frac{n!}{2!} + \frac{n!}{3!} - \cdots + \frac{n!}{(n-1)!} < 1.$$

But this completes the proof. Why? ■

It would make an interesting project to compare the proofs for the irrationality of the four numbers—

$$\sqrt{2}, \quad e, \quad \pi, \quad \text{and} \quad \sum_{n=1}^{\infty} \frac{1}{n^3}.$$

Problem 24. Subseries of the Harmonic Series

Calculus students usually have little intuitive feeling as to whether a given series of positive terms converges or diverges. The various convergence tests are of limited help, since they are seen as rules to manipulate formally without providing understanding. The concept of the gap sequence in this problem provides an intuitive idea to test a given series. Theorems 1 and 2 apply to most series encountered by calculus students. We consider a restricted collection of series to make the initial calculation easier, namely "subseries" of the harmonic series

$$\sum_{n=1}^{\infty} \frac{1}{n}.$$

That is, we consider series of the form

$$\sum_{n=1}^{\infty} \frac{1}{a_n}$$

where $a_1 = 1$ and $\{a_n\}$ is an increasing sequence of positive integers. Later we can permit $a_1 \neq 1$ and a_n to be non-integral. For each such series, we introduce the *gap sequence* $\{A_n\}$ where $A_n = a_{n+1} - a_n$. The name indicates that we are measuring the gap between successive denominators of the terms of the given series. For example, $\{1, 2, 3, 4, \ldots\}$ is the gap sequence for the series $1 + \frac{1}{2} + \frac{1}{4} + \frac{1}{7} + \frac{1}{11} + \cdots$ while $\{3, 5, 7, 9, \ldots\}$ is the gap sequence for the series $1 + \frac{1}{4} + \frac{1}{9} + \frac{1}{16} + \cdots$ and $\{2, 2, 2, 2, \ldots\}$ is the gap sequence for the series $1 + \frac{1}{3} + \frac{1}{5} + \frac{1}{7} + \cdots$.

In the following discussion, let

$$\sum_{n=1}^{\infty} \frac{1}{a_n} \quad \text{and} \quad \sum_{n=1}^{\infty} \frac{1}{b_n}$$

be two subseries of

$$\sum_{n=1}^{\infty} \frac{1}{n}$$

having the values of A and B respectively. If

$$\sum_{n=1}^{\infty} \frac{1}{a_n}$$

diverges, then $A = \infty$. Otherwise the series converges to A. Also, let $\{A_n\}$ and $\{B_n\}$ be the respective gap sequences for these two series.

Definition 1. $\{A_n\} \leq \{B_n\}$ means that $A_n \leq B_n$ for all n.

Lemma 1. *If* $\{A_n\} \leq \{B_n\}$, *then* $a_n \leq b_n$ *for all n and also*

$$\sum_{n=1}^{\infty} \frac{1}{a_n} \geq \sum_{n=1}^{\infty} \frac{1}{b_n}.$$

Proof We prove that $a_n \leq b_n$ by mathematical induction. It is trivially true for $n = 1$ because $a_1 = b_1 = 1$. Assume that $a_k \leq b_k$ is true. Supply reasons for the steps in the following proof.

$$\begin{aligned} a_{k+1} &= a_{k+1} - a_k + a_k \\ &= A_k + a_k \\ &\leq B_k + a_k \\ &= b_{k+1} - b_k + a_k \\ &= b_{k+1} - (b_k - a_k) \\ &\leq b_{k+1} \end{aligned}$$

■

Example 1. The series

$$\sum_{n=1}^{\infty} \frac{1}{b_n} = 1 + \frac{1}{2} + \frac{1}{4} + \frac{1}{7} + \frac{1}{11} + \cdots$$

converges. This follows by the limit form of the comparison test using

$$\sum_{n=1}^{\infty} \frac{1}{n^2}$$

as the comparison series because

$$\frac{1}{b_n} = \frac{1}{\frac{1}{2}n(n+1)+1} = \frac{2}{n^2 - n + 2}.$$

Notice that the gap sequence for this series is $\{B_n\} = \{1, 2, 3, 4, \ldots\}$.

Theorem 1. *If the series*

$$\sum_{n=1}^{\infty} \frac{1}{a_n}$$

has an increasing gap sequence $\{A_n\}$ *(i.e.,* $A_n < A_{n+1}$ *for all n), then the series*

$$\sum_{n=1}^{\infty} \frac{1}{a_n}$$

converges.

Proof Because $\{A_n\}$ is an increasing sequence, $\{A_n\} \geq \{B_n\}$ where $\{B_n\}$ is defined in Example 1 as the smallest increasing sequence. By Lemma 1,

$$\sum_{n=1}^{\infty} \frac{1}{a_n} \leq \sum_{n=1}^{\infty} \frac{1}{b_n},$$

and because the series

$$\sum_{n=1}^{\infty} \frac{1}{b_n}$$

converges, so does

$$\sum_{n=1}^{\infty} \frac{1}{a_n}.$$

■

Example 2. The constant gap sequence $\{D_n\} = \{k, k, k, k, \ldots\}$ is associated with the series

$$1 + \frac{1}{k+1} + \frac{1}{2k+1} + \frac{1}{3k+1} + \cdots$$

and this series diverges to ∞ by the integral test or the following argument.

$$\begin{aligned} 1 + \frac{1}{k+1} + \frac{1}{2k+1} + \frac{1}{3k+1} + \cdots &> 1 + \frac{1}{k+k} + \frac{1}{2k+k} + \frac{1}{3k+k} + \cdots \\ &= 1 + \frac{1}{2k} + \frac{1}{3k} + \cdots \\ &= 1 + \frac{1}{k}\left(\frac{1}{2} + \frac{1}{3} + \frac{1}{4} + \cdots\right) \\ &= \infty \end{aligned}$$

Theorem 2. *If the series*

$$\sum_{n=1}^{\infty} \frac{1}{a_n}$$

has a bounded gap sequence, then the series

$$\sum_{n=1}^{\infty} \frac{1}{a_n}$$

diverges to ∞.

Proof Suppose the gap sequence $\{A_n\}$ is bounded by k, in that $A_n \leq k$ for all n. Then $\{A_n\} \leq \{D_n\}$ as defined in Example 2. But by Lemma 1,

$$\sum_{n=1}^{\infty} \frac{1}{a_n} \geq \sum_{n=0}^{\infty} \frac{1}{nk+1}.$$

Therefore

$$\sum_{n=1}^{\infty} \frac{1}{a_n}$$

diverges to ∞. ■

The only subseries of

$$\sum_{n=1}^{\infty} \frac{1}{n}$$

we cannot handle by use of Theorems 1 and 2 are those whose gap sequence is unbounded and not strictly increasing. Some examples are given below in the exercises. Based on what you find in these exercises, make some conjectures about the convergence or divergence for this remaining category of series.

Exercises

1. Let the gap sequence be $\{1, 2, 1, 3, 1, 4, 1, 5, \ldots\}$, so that the related series is

$$1 + \tfrac{1}{2} + \tfrac{1}{4} + \tfrac{1}{5} + \tfrac{1}{8} + \tfrac{1}{9} + \tfrac{1}{13} + \tfrac{1}{14} + \tfrac{1}{19} + \cdots.$$

Notice that this series can be rewritten as the sum of the two series

$$1 + \tfrac{1}{4} + \tfrac{1}{8} + \tfrac{1}{13} + \cdots \quad \text{and} \quad \tfrac{1}{2} + \tfrac{1}{5} + \tfrac{1}{9} + \tfrac{1}{14} + \cdots,$$

each of which converges by Theorem 1. Another way to show that this series converges is to observe that

$$\left(1 + \frac{1}{2}\right) + \left(\frac{1}{4} + \frac{1}{5}\right) + \left(\frac{1}{8} + \frac{1}{9}\right) + \cdots < 2 \cdot 1 + 2 \cdot \frac{1}{4} + 2 \cdot \frac{1}{8} + 2 \cdot \frac{1}{13} + \cdots$$

$$= \sum_{n=1}^{\infty} \frac{2}{\frac{1}{2}n^2 + \frac{3}{2}n - 1}.$$

Supply reasons for the statements made above.

2. Let the gap sequence be $\{1, 2, 2, 3, 3, 3, 4, 4, 4, 4, \ldots\}$, so that the related series is

$$1 + \tfrac{1}{2} + \tfrac{1}{4} + \tfrac{1}{6} + \tfrac{1}{9} + \tfrac{1}{12} + \tfrac{1}{15} + \tfrac{1}{19} + \tfrac{1}{23} + \tfrac{1}{27} + \tfrac{1}{31} + \cdots.$$

Verify the following result.

$$1+\frac{1}{2}+\left(\frac{1}{4}+\frac{1}{6}\right)+\left(\frac{1}{9}+\frac{1}{12}+\frac{1}{15}\right)+\cdots<1+\frac{1}{2}+2\cdot\frac{1}{4}+3\cdot\frac{1}{9}+4\cdot\frac{1}{19}$$
$$+\cdots=1+\frac{1}{2}+\sum_{n=2}^{\infty}\frac{n}{\frac{1}{3}n^3-\frac{1}{2}n^2+\frac{7}{6}n+1}$$

Show that the denominator on the right side of the inequality above is equal to the expression $2+2^2+3^2+\cdots+(n-1)^2+n$. Also explain why the above information shows the given series converges.

3. Let the gap sequence be $\{1, 2, 1, 1, 3, 1, 1, 1, 4, 1, 1, 1, 1, 5, \ldots\}$, so that the related series is

$$1+\tfrac{1}{2}+\tfrac{1}{4}+\tfrac{1}{5}+\tfrac{1}{6}+\tfrac{1}{9}+\tfrac{1}{10}+\cdots.$$

Verify the following inequality and show it implies the given series diverges to infinity.

$$\left(1+\tfrac{1}{2}\right)+\left(\tfrac{1}{4}+\tfrac{1}{5}+\tfrac{1}{6}\right)+\left(\tfrac{1}{9}+\cdots+\tfrac{1}{12}\right)+\left(\tfrac{1}{16}+\cdots+\tfrac{1}{20}\right)+\cdots$$
$$>\tfrac{2}{2}+\tfrac{3}{6}+\tfrac{4}{12}+\tfrac{5}{20}+\cdots$$

4. Let the gap sequence be $\{1, 1, 2, 2, 4, 2, 4, 2, 4, 6, 2, 6, \ldots\}$. This unusual sequence is related to a very famous series, namely that of the reciprocals of the prime numbers.

$$1+\tfrac{1}{2}+\tfrac{1}{3}+\tfrac{1}{5}+\tfrac{1}{7}+\tfrac{1}{11}+\tfrac{1}{13}+\tfrac{1}{17}+\tfrac{1}{19}+\tfrac{1}{23}+\tfrac{1}{29}+\tfrac{1}{31}+\tfrac{1}{37}+\cdots$$

This sequence is unbounded because there can be arbitrarily large runs of consecutive composite numbers. The twin prime conjecture (which is still unproven) asserts that the value 2 occurs infinitely often as a term in this sequence. It is known that this series diverges—one proof of this was given by Euler in 1737.[19] Verify the statements made in this exercise.

5. Based on your findings in Exercises 1–4 and other examples you consider, make a conjecture about which unbounded and non-increasing gap sequences lead to a convergent series. Then try to prove your conjecture in order to obtain a Theorem 3 to extend the results given above in Theorems 1 and 2.

Problem 25. Finding Functions whose Maclaurin Series is $\sum_{n=1}^{\infty} n^k x^n$

A non-traditional way to view a Maclaurin series representation of a function is as an element of a vector space over the reals, whose basis consists of the infinite collection $\{1, x, x^2, \ldots, x^n, \ldots\}$. An arbitrary element in this space is the linear combination $\sum_{n=0}^{\infty} a_n x^n$, where a_n is a rational number for every example seen in calculus. The usual task is to start with a known function $f(x)$ and ask how to find the vector space element associated with it in the sense of Taylor's theorem (see Problem 3). The well-known answer

[19]See pp. 70–6 in [**10**].

is that

$$a_n = \frac{f^{(n)}(0)}{n!}$$

if $f(x)$ satisfies certain properties. The reverse question can be more difficult. That is, if we are given a member $\sum_{n=0}^{\infty} a_n x^n$ of this vector space, find the function $f(x)$ so that the Maclaurin series representation of $f(x)$ is the given expression $\sum_{n=0}^{\infty} a_n x^n$.

Example 1. The following example shows an application of this reverse process. Suppose we wanted to find the exact value for the series

$$\sum_{n=1}^{\infty} \frac{n}{3^n}.$$

We know by the ratio test that this series converges, but this gives no idea of its exact value. Consider the power series $\sum_{n=1}^{\infty} nx^n$. If we can find the function $f(x)$ whose Maclaurin series is given by $\sum_{n=1}^{\infty} nx^n$, then the value of the series

$$\sum_{n=1}^{\infty} \frac{n}{3^n}$$

will be $f(\frac{1}{3})$. As we will see in Example 2,

$$f(x) = \frac{x}{(1-x)^2},$$

so the given series converges to

$$f\left(\tfrac{1}{3}\right) = \tfrac{3}{4}.$$

We use the geometric series result next to determine a recursive formula for the function whose Maclaurin series is given by $\sum_{n=1}^{\infty} n^k x^n$ for any non-negative integer k. We observe first that when $k = 0$, we get the geometric series $\sum_{n=1}^{\infty} x^n$, which is known to converge to $x/(1-x)$ whenever $|x| < 1$.

Example 2. For the case when $k = 1$, we want to find the function $f(x)$ whose Maclaurin series is $\sum_{n=1}^{\infty} nx^n$. The following steps yield a solution.

$$f(x) = \sum_{n=1}^{\infty} nx^n = x + 2x^2 + 3x^3 + \cdots$$

$$\frac{f(x)}{x} = \sum_{n=1}^{\infty} nx^{n-1} = 1 + 2x + 3x^2 + \cdots$$

$$\int \frac{f(x)}{x}\,dx = \sum_{n=1}^{\infty} x^n = x + x^2 + x^3 + \cdots = \frac{x}{1-x}$$

by the case when $k = 0$, so

$$\frac{f(x)}{x} = \frac{d}{dx}\left(\frac{x}{1-x}\right) = \frac{1}{(1-x)^2}$$

$$f(x) = \sum_{n=1}^{\infty} nx^n = \frac{x}{(1-x)^2}$$

Example 3. For the case when $k = 2$, we carry the work just far enough to see the recursive pattern.

$$f(x) = \sum_{n=1}^{\infty} n^2 x^n = x + 4x^2 + 9x^3 + 16x^4 + \cdots$$

$$\frac{f(x)}{x} = \sum_{n=1}^{\infty} n^2 x^{n-1} = 1 + 4x + 9x^2 + 16x^3 + \cdots$$

$$\int \frac{f(x)}{x}\,dx = \sum_{n=1}^{\infty} nx^n = \frac{x}{(1-x)^2}$$

using the result in Example 2. Therefore

$$f(x) = \sum_{n=1}^{\infty} n^2 x^n = x\,\frac{d}{dx}\left(\frac{x}{(1-x)^2}\right) = \frac{x+x^2}{(1-x)^3}$$

In this problem,

$$\sum_{n=1}^{\infty} n^2 x^n = x\,\frac{d}{dx}\left(\sum_{n=1}^{\infty} nx^n\right).$$

This result can be generalized to

$$\sum_{n=1}^{\infty} n^k x^n = x\,\frac{d}{dx}\left(\sum_{n=1}^{\infty} n^{k-1} x^n\right).$$

You are asked to prove this result in Exercise 1 and then to apply it in Exercises 2–4.

Example 4. The recursion formula given in Exercise 1 applies to negative integer values of k also, but they give a less satisfying outcome than positive integer values of k. For the case when $k = -1$, we have

$$\sum_{n=1}^{\infty} n^{-1} x^n = x\,\frac{d}{dx}\left(\sum_{n=1}^{\infty} n^{-2} x^n\right).$$

By rewriting this equation and substituting in the function value for

$$\sum_{n=1}^{\infty} \frac{x^n}{n}$$

from Exercise 4, we have

$$\sum_{n=1}^{\infty} \frac{x^n}{n^2} = \int \frac{\sum_{n=1}^{\infty} \frac{x^n}{n}}{x}\, dx = \int \frac{-\ln(1-x)}{x}\, dx.$$

We know we cannot simplify the integral in Example 4 because otherwise we would have a closed form expression for

$$f(x) = \sum_{n=1}^{\infty} \frac{x^2}{n^2},$$

which means that $f(1)$ should give us the value of the p-series

$$\sum_{n=1}^{\infty} \frac{1}{n^2}.$$

This would be much too easy a solution for this well-known problem (see Problem 11). In Exercise 5, we begin to see that

$$\int \frac{-\ln(1-x)}{x}\, dx$$

may not have a simple antiderivative.

Exercises

1. Verify the recursion formula that

$$\sum_{n=1}^{\infty} n^k x^n = x \frac{d}{dx} \left(\sum_{n=1}^{\infty} n^{k-1} x^n \right).$$

2. Verify that

$$\sum_{n=1}^{\infty} n^3 x^n = \frac{x + 4x^2 + x^3}{(1-x)^4}.$$

Use this formula to find the exact value for the series

$$\sum_{n=1}^{\infty} \frac{n^3}{2^n}$$

and for the series

$$\sum_{n=1}^{\infty} \frac{n^3}{3^n}.$$

3. Verify that

$$\sum_{n=1}^{\infty} n^4 x^n = \frac{x + 11x^2 + 11x^3 + x^4}{(1-x)^5}.$$

4. Let $f(x)$ be the function whose Maclaurin series is

$$\sum_{n=1}^{\infty} \frac{x^n}{n}.$$

Use the geometric series idea as in Example 2, or use the recursion formula when $k = 0$ in Exercise 1, to show that $f(x) = -\ln|1 - x|$.

5. Use the substitution $u = \ln(1 - x)$ to change

$$\int \frac{-\ln(1-x)}{x}\,dx$$

to the equivalent integral

$$\int \frac{u}{e^{-u} - 1}\,du.$$

These are two examples of integrals for which there is no simple antiderivative.

Problem 26. A Pascal-like Triangle

In their book **[8]**, John Conway and Richard Guy present many examples of interesting patterns of numbers. This problem deals with a pattern that is in the spirit of those found in this book. It is based on Problem 25 where we found the first several functions $I_k(x)$ whose Maclaurin series representation is given by $\sum_{n=1}^{\infty} n^k x^n$. The beginning terms of this sequence are listed below.

$$I_0(x) = \sum_{n=1}^{\infty} x^n = \frac{x}{(1-x)}$$

$$I_1(x) = \sum_{n=1}^{\infty} n x^n = \frac{x}{(1-x)^2}$$

$$I_2(x) = \sum_{n=1}^{\infty} n^2 x^n = \frac{x + x^2}{(1-x)^3}$$

$$I_3(x) = \sum_{n=1}^{\infty} n^3 x^n = \frac{x + 4x^2 + x^3}{(1-x)^4}$$

$$I_4(x) = \sum_{n=1}^{\infty} n^4 x^n = \frac{x + 11x^2 + 11x^3 + x^4}{(1-x)^5}$$

$$I_5(x) = \sum_{n=1}^{\infty} n^5 x^n = \frac{x + 26x^2 + 66x^3 + 26x^4 + x^5}{(1-x)^6}$$

$$I_6(x) = \sum_{n=1}^{\infty} n^6 x^n = \frac{x + 57x^2 + 302x^3 + 302x^4 + 57x^5 + x^6}{(1-x)^7}$$

It seems reasonable from these functions to assume we can express $I_k(x)$ in the following form.

Conjecture 1. *$I_k(x)$ can be written in the form*

$$I_k(x) = \frac{\sum_{j=1}^{k} a_j x^j}{(1-x)^{k+1}} \quad \textit{for some positive integers } a_j.$$

Proof We prove this conjecture using mathematical induction. Clearly the result is true for $k = 1$, so we next assume it is true for $I_k(x)$. We also remember the source of these functions from Problem 25 and use the fact that

$$I_{k+1}(x) = x \frac{d}{dx} I_k(x).$$

Therefore

$$\begin{aligned}
I_{k+1}(x) &= x \frac{d}{dx} I_k(x) \\
&= x \frac{d}{dx} \left(\frac{\sum_{j=1}^{k} a_j x^j}{(1-x)^{k+1}} \right) \\
&= x \left(\frac{(1-x)^{k+1} \left(\sum_{j=1}^{k} j a_j x^{j-1} \right) - \left(\sum_{j=1}^{k} a_j x^j \right) (k+1)(1-x)^k(-1)}{(1-x)^{2k+2}} \right) \\
&= x \left(\frac{(1-x) \sum_{j=1}^{k} j a_j x^{j-1} + (k+1) \sum_{j=1}^{k} a_j x^j}{(1-x)^{k+2}} \right) \\
&= \frac{\sum_{j=1}^{k} j a_j x^j - \sum_{j=1}^{k} j a_j x^{j+1} + \sum_{j=1}^{k} (k+1) a_j x^{j+1}}{(1-x)^{k+2}} \\
&= \frac{\sum_{j=1}^{k} j a_j x^j - \sum_{j=2}^{k+1} (j-1) a_{j-1} x^j + \sum_{j=2}^{k+1} (k+1) a_{j-1} x^j}{(1-x)^{k+2}} \\
&= \frac{(a_1)x + \sum_{j=2}^{k} (j a_j + (k-j+2) a_{j-1}) x^j + (-k a_k + (k+1) a_k) x^{k+1}}{(1-x)^{k+2}} \\
&= \frac{\sum_{j=1}^{k+1} b_j x^j}{(1-x)^{k+2}} \quad \text{for some positive integers } b_j.
\end{aligned}$$

■

We next look at the array of numbers determined by these coefficients a_j of the x^j term. It is useful to use the double-subscripted symbol $A_{k,j}$ to represent the jth element in the kth row. If we replace a_j by $A_{k,j}$ and replace b_j by $A_{k+1,j}$ in our proof of Conjecture 1, we can represent the general function by

$$I_k(x) = \frac{\sum_{j=1}^{k} A_{k,j} x^j}{(1-x)^{k+1}}$$

and have the following recursion formulas

$$A_{k+1,1} = A_{k,1}$$
$$A_{k+1,j} = (k-j+2)A_{k,j-1} + jA_{k,j} \quad \text{for } 2 \leq j \leq k$$
$$A_{k+1,k+1} = A_{k,k}.$$

The triangular array of the $A_{k,j}$ numbers below remind us of Pascal's triangle. By looking at the first six rows of the triangle and recalling the properties of Pascal's triangle, we are led to make the following conjectures.

```
                1
              1   1
            1   4   1
         1   11   11   1
      1   26   66   26   1
   1   57   302   302   57   1
```

Conjecture 2. $A_{k,1} = A_{k,k} = 1$ *(i.e., Each row begins and ends with the number 1).*

Conjecture 3. $A_{k,j} = A_i k, k+1-j$ *(i.e., Each row has a right-to-left symmetry).*

Conjecture 4. *From the calculations in Exercise 1, we propose that the second value in the kth row is given by*

$$A_{k,2} = 2A_{k-1,2} + (k-1)\,2^k - (k+1).$$

In Pascal's triangle, we know that each entry is equal to the sum of the two elements directly above it. This leads us to ask if the general element of our new triangle is equal to some linear combination of the two elements above it.

Conjecture 5. $A_{k,j} = \alpha A_{k-1,j-1} + \beta A_{k-1,j}$ *for some values of α and β.*

If we remember that the sum of the entries in the kth row of Pascal's triangle sums to 2^k, we might consider the sum of the entries in the kth row of this new triangle. The values of $1, 2, 6, 24, 120, 720, \ldots$ lead us to the next conjecture.

Conjecture 6. $\sum_{j=1}^{k} A_{k,j} = k!$.

It turns out all these conjectures are true. Some proofs are given below and others are left as exercises.

Proof of Conjecture 3 $A_{k,j} = A_{k,k-j+1}$.
This result is clearly true for the first six values of k. We next assume it is true for the general case k, and prove it is true for the $(k+1)$st case, namely that

$$A_{k+1,j} = A_{k+1,k-j+2}.$$

We only need to consider when $2 \le j \le k$ since the cases when $j = 1$ and $j = k + 1$ are handled by Conjecture 2. For these values, we have

$$A_{k+1,j} = jA_{k,j} + (k - j + 2)A_{k,j-1} \qquad \text{and so}$$
$$A_{k+1,k-j+2} = (k - j + 2)A_{k,k-j+2} + (k - (k - j + 2) + 2)A_{k,k-j+1}. \qquad \blacksquare$$

You are asked in Exercise 3 to complete this proof of Conjecture 3. Observe that Conjecture 5 has already been proven as a consequence of the proof of Conjecture 1, and this result was also used in the proof of Conjecture 3. ■

We are ready for the proof of Conjecture 6 that $\sum_{j=1}^{k} A_{k,j} = k!$.

Proof of Conjecture 6 We observe this conjecture is true for the first six values of k. We assume the induction hypothesis that $\sum_{j=1}^{k} A_{k,j} = k!$, and prove that $\sum_{j=1}^{k+1} A_{k+1,j} = (k + 1)!$. It helps to arrange the terms in a columnar form, and then add terms along the diagonal.

$$\begin{aligned}
\sum_{j=1}^{k+1} A_{k+1,j} &= A_{k+1,1} + \sum_{j=2}^{k} A_{k+1,j} + A_{k+1,k+1} \\
&= A_{k,1} \\
&\quad + (2A_{k,2} + kA_{k,1}) \\
&\quad + (3A_{k,3} + (k-1)A_{k,2}) \\
&\quad + (4A_{k,4} + (k-2)A_{k,3}) \\
&\quad \vdots \\
&\quad + ((k-1)A_{k,k-1} + 3A_{k,k-2}) \\
&\quad + (kA_{k,k} + 2A_{k,k-1}) \\
&\quad + A_{k,k} \\
&= (k+1)\sum_{j=1}^{k} A_{k,j} = (k+1)k! = (k+1)!
\end{aligned} \qquad \blacksquare$$

It is quite exciting that a Pascal-like triangle should emerge in such a natural, yet unexpected, way as it does in this problem. We wonder if there are other settings where the solution can be expressed in such a triangular pattern. The harmonic triangle of Leibniz is based on a pattern of subtracting elements, instead of adding them. It offers a nice contrast and comparison to this problem. A recent article[20] presents an alternate method for finding the $A_{k,j}$ numbers and identifies them as *Eulerian numbers* (see Exercise 6).

[20] James A. Sellers, "Beyond Mere Convergence," *PRIMUS*, Volume XII, Number 2, June 2002, pp. 157–164.

Exercises

1. Verify that

$$x\,\frac{d}{dx}\left(\frac{x+ax^2+x^3}{(1-x)^4}\right)=\frac{x+(2a+3)x^2+(2a+3)x^3+x^4}{(1-x)^5}$$

and that

$$x\,\frac{d}{dx}\left(\frac{x+bx^2+bx^3+x^4}{(1-x)^5}\right)=\frac{x+(2b+4)x^2+6bx^3+(2b+4)x^4+x^5}{(1-x)^6}.$$

These calculations support Conjecture 4.

2. Prove Conjecture 2 that $A_{k,1}=1$ and $A_{k,k}=1$.

3. Complete the proof of Conjecture 3.

4. Verify that Conjecture 4 is a special case of Conjecture 5.

5. Use the recursion formulas to verify the correctness of the elements in the 5th and 6th rows of the triangle above. Then find the values in the 7th row.

6. There is one final conjecture of interest. In Pascal's triangle, we not only have a recursive formula for the jth element in the kth row, we also have an explicit formula, namely

$$C(k,j)=\binom{k}{j}=\frac{k!}{j!(k-j)!}.$$

The formula

$$A_{k,j}=\sum_{i=0}^{j}(-1)^i(j-i)^k\binom{k+1}{i}$$

is proposed at an online site.[21] Verify the correctness of this formula for some specific cases.

Problem 27. Finding Areas without Antiderivatives

Measurement has always been an important topic in mathematics. The word 'geometry' refers to measurement of the earth and this task was carried out in ancient cultures such as the Egyptians. The Greeks made geometry a more precise science through the writings of Euclid about 300 B.C. In this problem, we consider two different approaches that were used in the ancient Greek period to find areas of various regions.

[21] N.J.A. Sloane, *The On-Line Encyclopedia of Integer Sequences*. Published electronically as `http://www.research.att.com/~njas/sequences/`.

I. The Early Approach of the Greeks to the Finding of Area The Greeks had a strong interest in the measurement of area. Such simple figures as the rectangle and right triangle received much of the early attention, and then interest was extended to more complicated shapes such as those of the circle and the parabola. Greek mathematicians tried to show the area of a more complicated shape was exactly equal to the area of a related right triangle or rectangle, whose area was known.

The trisection of an angle is one of three famous construction problems posed by the Greeks, which received much attention from outstanding mathematicians over many centuries until these constructions were shown in the mid 1800s to be impossible to accomplish according to the rules they imposed. A second of these construction problems is that of squaring the circle. In this problem, you are given a circle (which is considered a more complicated shape) and asked to construct a square (which is considered a much simpler shape) that has the same area.

The first success of starting with a more complicated shape and constructing a simpler shape having the same area was achieved by Hippocrates about 450 B.C. It is known today as the *quadrature of the lune*. Quadrature is an older term that refers to finding the area of a region.

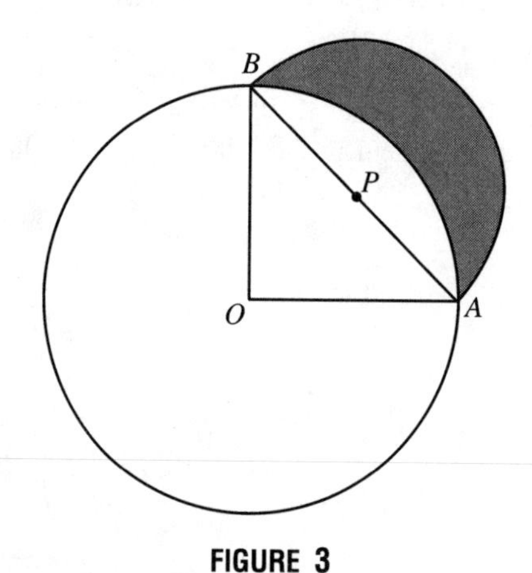

FIGURE 3

We begin with the right triangle AOB that is inscribed in a circle of radius a (see Figure 3). A portion of a second circle is then drawn, whose center is at point P (the midpoint of line segment AB) and whose radius is equal to AP. The portion of this second circle lying outside the first circle can be shown to have the same area as the triangle AOB. Since this shape resembles the crescent of the moon, it has been called a lune, or a lunar crescent.

The Greeks tried to find areas of other regions, such as the quadrature of the circle and the quadrature of the parabola, as shown in Figure 4. A straight line cuts off a piece from a circle or a parabola, and the problem is to find the area of that piece. No one was able to solve the quadrature of the circle problem in the Greek period, but Archimedes was able to solve the quadrature of the parabola problem by use of an infinite geometric series. By choosing the point P on the parabola where the tangent line AB is parallel to the given line,

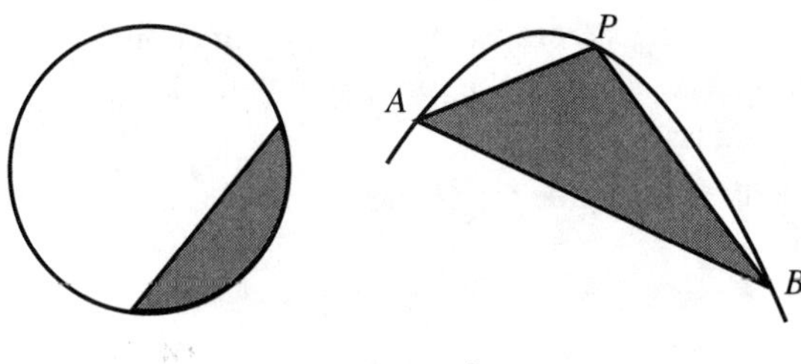

FIGURE 4

and using this point as the vertex of an inscribed triangle, Archimedes proved the area of the parabolic segment was equal to $\frac{4}{3}$ times the area of the triangle.[22]

In another problem, Archimedes proved before 200 B.C. that the area of a circle with radius r and circumference C is equal to the area of a right triangle having one leg of length r and the other leg of length C. There was at this time no such formula as πr^2 for the area of a circle. Archimedes' method was to assume the area of the circle was less than the area of the triangle described above, and then find a contradiction. Similarly, he assumed the area of the circle was greater than the area of the triangle and found another contradiction. The only option left was that the two areas were equal.[23] This technique is called the *method of exhaustion*.

II. The Later Approach of the Greeks to Area Archimedes was also the first one to tackle the problem of finding a value for our modern symbol of π. He did this by a complete shift in method. Instead of using the traditional Greek approach of showing that the area of a complex region was equal to the area of a well-known simpler shape, he developed the new method of approximation.

It first appeared in his attempt to find the area of a circle with radius 1. His idea was to inscribe a regular hexagon inside the circle. Since the hexagon consists of six equilaterial triangles of side length 1 (or 12 right triangles), the area of the hexagon is

$$\frac{3\sqrt{3}}{2} \approx 2.598,$$

which is an approximation (a lower bound) for the area of the circle which is π. In addition, the perimeter of the hexagon is 6, which is an approximation for the circumference of the circle, which is 2π. By either means Archimedes could approximate the value for π. The perimeter approach is perhaps the easier one. We first consider a general principle about approximations.

A General Method for Successful Approximations

1. Clearly state what quantity Q you want to find.
2. Determine a method for finding an approximation for Q. It is most desirable to use a method with the property that for any approximation, it is possible to repeat the method and find a better approximation.

[22] See pp. 310–14 in [**31**].
[23] See pp. 92–6 in [**11**].

3. Find a formula for the general case of the approximation. That is, find a formula for the nth approximation Q_n with the property that the accuracy of the approximation increases as n gets larger.
4. If possible, find the limit value for Q_n, and define

$$Q = \lim_{n\to\infty} Q_n.$$

Archimedes' method met the criteria of step 2, since given his approximation of π using the hexagon, he could next inscribe a regular 12-sided polygon inside the circle and obtain a better approximation (see Figure 5). This process of doubling can be repeated indefinitely. He developed the recursion formula that if s_n is the length of one side of the regular n-sided inscribed polygon, then the length of one side of the regular $2n$-sided inscribed polygon will be

$$s_{2n} = \sqrt{2 - \sqrt{4 - s_n^2}}.$$

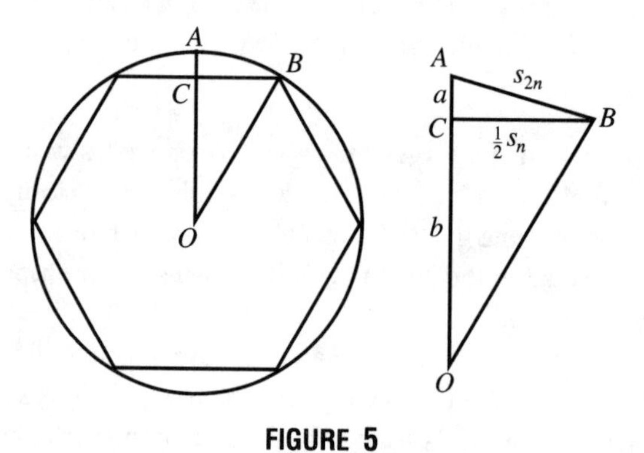

FIGURE 5

In Exercise 3 you are asked to prove this recursion formula by considering Figure 5 and the three equations given below.

$$a^2 + \tfrac{1}{4}s_n^2 = s_{2n}^2 \tag{1}$$

$$b^2 + \tfrac{1}{4}s_n^2 = 1 \tag{2}$$

$$a + b = 1 \tag{3}$$

If P_n represents the perimeter of the regular n-sided polygon, then $P_n = n\,s_n$, and therefore

$$\pi \approx \frac{P_n}{2} = \frac{n}{2}s_n.$$

We show next that the area of the regular n-sided polygon is given by

$$A_{2n} = \frac{n}{2}s_n,$$

for values of $n = 6$, 12, 24, and 48. Notice also that $A_{2n} = \frac{1}{2} P_n$.

$$\begin{aligned} A_{2n} &= 2n(\text{area of } \triangle AOB) \\ &= 2n(\text{area of } \triangle ACB) + 2n(\text{area of } \triangle OCB) \\ &= 2n\left(\frac{1}{4} a s_n\right) + 2n\left(\frac{1}{4} b s_n\right) \\ &= \frac{n}{2} s_n (a+b) = \frac{n}{2} s_n. \end{aligned}$$

Archimedes calculated s_n, P_n and A_n for the values of n shown in the table below. As you can see, although a significant amount of calculation was required considering the tools available around 200 B.C. for this purpose, only three decimal places of accuracy were achieved.

n	s_n	$\frac{P_n}{2}$	A_n
6	1	3.0000	$\frac{3\sqrt{3}}{2} \approx 2.5980$
12	$\sqrt{2-\sqrt{3}} \approx .5176$	3.1058	$3s_6 = 3.0000$
24	$\sqrt{2-\sqrt{2+\sqrt{3}}} \approx .2610$	3.1326	$6s_{12} \approx 3.1058$
48	$\sqrt{2-\sqrt{2+\sqrt{2+\sqrt{3}}}} \approx .1308$	3.1393	$12s_{24} \approx 3.1326$
96	$\sqrt{2-\sqrt{2+\sqrt{2+\sqrt{2+\sqrt{3}}}}} \approx .06543$	3.1410	$24s_{48} \approx 3.1393$

Archimedes quit after using a polygon of 96 sides, probably because of the complexity of the calculations. This did not deter later mathematicians, such as Vieté who used a polygon of $6 \cdot 2^{16} = 393{,}216$ sides in 1579 A.D. to approximate π to nine decimal places. Another used a polygon with $2^{62} > 4 \cdot 10^{18}$ sides to approximate π correctly to 35 decimal places in 1610. We show in Problem 14 that the infinite Maclaurin series expansion of $\arctan x$ provides a simpler method for approximations of π, which have reached into the billions of decimal place accuracy today with the use of computers.

Clearly the method of Archimedes fails at step 4 of the general method for approximations stated above. After Archimedes, no one came close to the level of his work on areas and volumes for almost 2000 years. After the Renaissance in western Europe reawakened interest in academic matters in general, a renewed interest in science surfaced in the late 1500s and early 1600s, especially in the field of astronomy. Such men as Copernicus, Kepler and Galileo initiated a study of planetary observation, which they broadened to include motion of all kinds and ways to find areas of more complicated regions than the ones considered by Archimedes. See Essay 1 also.

We see in Problems 4 and 16 how Cavalieri and Fermat developed summation formulas using areas of rectangles instead of polygons to find the area of the region under $y = x^k$ and above the interval [0, 1]. The method in Problem 4 yields a recursion formula, while Fermat's method in Problem 16 was more successful in that both steps 3 and 4 could be applied.

Exercises

1. Use modern formulas to verify the shaded area of the lune in Figure 3 equals the area of the right triangle BOA.
2. By using the modern method of integration, show that a solution for the quadrature of the circle problem (see Figure 4) can be expressed by A, which gives the area for the part of the circle $x^2 + y^2 = a^2$ that is cut off by the straight line $y = mx + b$,

$$A = \int_{x_1}^{x_2} \left| \sqrt{a^2 - x^2} - mx - b \right| dx$$

where the x-coordinates of the points of intersection of the circle and the line are given by

$$x_1 = \frac{-mb - \sqrt{m^2a^2 + a^2 - b^2}}{1 + m^2},$$

$$x_2 = \frac{-mb + \sqrt{m^2a^2 + a^2 - b^2}}{1 + m^2}.$$

It is not necessary to evaluate this integral.
3. Obtain the recursion formula

$$s_{2n} = \sqrt{2 - \sqrt{4 - s_n^2}}$$

by solving the system of three equations after Figure 5.
4. Verify the values in the table for s_{24} and s_{48}.
5. Give reasons for the proof that

$$A_{2n} = \frac{n}{2} s_n.$$

Problem 28. Questions from the Euler–Stirling Correspondence

The reading selection from the book *James Stirling* by Ian Tweddle (see Selection 2 in Part IV) consists of three letters exchanged between Leonhard Euler and James Stirling. Euler initiated the interchange in 1729 by sending a letter from St. Petersburg, where he had recently come in 1727, to the older and more established James Stirling in Scotland. Euler's hope was to gain recognition by members of the Royal Society in London for himself and his work. After two years, Stirling sent a reply, much to Euler's delight, who then immediately sent a second letter. As far as we know, this was the end of the correspondence, and the two individuals never met.

In Euler's letters especially, one gains insight into the human qualities of Euler and his excitement over the discovery of new ideas. These include the development of what is now called the Euler–Maclaurin formula, Euler's constant, and Euler's formulas for the value of the convergent p-series

$$\sum_{n=1}^{\infty} \frac{1}{n^{2k}}$$

in terms of the product of a rational number and π^{2k}. It is especially interesting to look for patterns in the coefficients of these formulas. What we now call the Bernoulli numbers are involved, but Euler gives no indication that he is aware of them at this time. He corrects an error in one term of the Euler–Maclaurin formula from his first letter and adds an additional term to the summation in the second letter.

We observe many interesting things, such as Euler's tongue-in-cheek observation of the weakness of Newton's fluxion notation, the publication requirements placed upon Euler by the St. Petersburg Academy, Stirling's defense of his countryman Colin Maclaurin and his role in developing the Euler–Maclaurin formula, and Euler's use of many interesting examples to illustrate his results. One can learn much by working through these examples. In doing so, one feels a sense of awe when observing Euler's regular use of 15–20 decimal place accuracy at a time when computers and hand-held calculators were not available. The following problems are based on this reading, and offer an excellent opportunity to recreate some of Euler's discoveries.

Exercises

Note: All equation numbers in these exercises refer to the numbered equations in the correspondence between Euler and Stirling in Selection 2 from Part IV.

1. Use a computer algebra system such as Derive, Maple, or Mathematica to verify the values for

$$\sum_{n=1}^{1000} \frac{1}{n} \quad \text{and} \quad \sum_{n=1}^{10^6} \frac{1}{n}$$

that are given in equation (2) in the correspondence between Euler and Stirling. How do you think Euler got these values to 16 decimal places with the calculating devices that were available to him in the mid-1700s?

2. Use the Euler–Maclaurin formula in equation (1) to obtain the formula for $\sum_{i=1}^{n} i^6$. When Cavalieri obtained this formula using a more complicated recursive method in the early 1600s (see Problem 4), he found that

$$\sum_{i=1}^{n} i^6 = \frac{n}{42}(n+1)(2n+1)(3n^4+6n^3-3n+1).$$

Show that your answer can be factored into this form.

3. Use the Euler–Maclaurin formula in equation (1) to obtain the formula for

$$1+\frac{1}{4}+\frac{1}{9}+\cdots+\frac{1}{x^2}$$

that is given in equation (3). Also find the term after the last one that Euler gave (which was $-5/66x^{11}$).

4. Use $x = 10$ and the formula obtained in Exercise 3 to find an approximation for the constant C, and compare with the value that Euler gave. Explain how the value for C is connected to the series

$$1+\frac{1}{2^2}+\frac{1}{3^2}+\frac{1}{4^2}+\frac{1}{5^2}+\text{etc.}$$

that is given in equation (4).

5. Rewrite the formula in equation (5) using modern notation. Use the Bernoulli numbers $\{B_n\}$ that are discussed in Problem 30. Repeat this exercise for the value of

$$\sum_{n=1}^{\infty} \frac{1}{n^{2k}}$$

in the formulas beginning with equation (7).

6. Show how to use Euler's result in equation (6) that

$$\frac{p^2}{8} = 1+\frac{1}{3^2}+\frac{1}{5^2}+\cdots$$

in order to establish the result that

$$\frac{p^2}{6} = 1+\frac{1}{2^2}+\frac{1}{3^2}+\frac{1}{4^2}+\cdots.$$

Problem 29. The Gamma Function

Functions defined by an integral play an important role in analysis. The first example encountered by a calculus student is likely the natural logarithm function, defined as

$$\ln x = \int_1^x \frac{1}{t}\,dt.$$

Many other functions can be defined using this general form of

$$F(x) = \int_a^x f(t)\,dt,$$

such as

$$\arctan x = \int_0^x \frac{1}{1+t^2}\, dt.$$

A second category of functions defined by an integral consists of those with the form $G(x) = \int_a^b g(x,t)\, dt$. The best known example is the gamma function, defined as

$$\Gamma(x) = \int_0^\infty t^{x-1} e^{-t}\, dt.$$

A third category consists of functions having the form

$$H(x,y) = \int_a^b h(x,y,t)\, dt.$$

One example is the beta function, defined by

$$\beta(x,y) = \int_0^1 t^{x-1}(1-t)^{y-1}\, dt.$$

The gamma and beta functions are called *special functions*, referring to a collection of important functions beyond the usual transcendental functions studied in calculus.

The gamma function was likely first studied by Euler, and is usually the first special function to be considered. It frequently appears as an optional calculus problem in the chapter on improper integrals. The first step in the examination of the function

$$\Gamma(x) = \int_0^\infty t^{x-1} e^{-t}\, dt$$

is to show that this integral converges for all $x > 0$. We do this by the following comments.

We begin by noting that when $x \geq 1$, the function $f(t) = t^{x-1}e^{-t}$ is continuous and hence integrable for all $t \geq 0$, while for $x < 1$, the function $f(t) = t^{x-1}e^{-t}$ has an infinite discontinuity at $t = 0$.

We also observe there are comparison tests for the convergence of improper integrals, just as there are for the convergence of infinite series. Here is the statement of one such test, which is used in Example 2.

Comparison test 1. If $\int_a^\infty g(x)\, dx$ converges to A and $0 \leq f(x) \leq g(x)$ for all $x \geq a$, then $\int_a^\infty f(x)\, dx$ converges also and to a value less than or equal to A. We assume $\int_a^\infty f(x)\, dx$ and $\int_a^\infty g(x)\, dx$ are improper integrals of the first kind.

Example 1. If $0 < x < 1$, then $\int_0^1 t^{x-1}e^{-t}\, dt$ is a convergent improper integral of the second kind.[24]

[24] If $f(x)$ is integrable on intervals of the form $[t,b]$ for $a < t \leq b$ and $\lim_{x\to a^+} |f(x)| = \infty$, then $\int_a^b f(x)\, dx$ is an *improper integral of the second kind* and $\int_a^b f(x)\, dx = \lim_{t\to a^+} \int_t^b f(x)\, dx$.

Solution. For $t \geq 0$, we have $0 < e^{-t} \leq 1$, and so $0 < t^{x-1}e^{-t} \leq t^{x-1}$. By Exercise 1, $\int_0^1 t^{x-1}\,dt$ converges if $0 < 1 - x < 1$, which holds for values of x for which $0 < x < 1$. Therefore $\int_0^1 t^{x-1}e^{-t}\,dt$ converges by comparison with $\int_0^1 t^{x-1}\,dt$. ■

Example 2. If $x > 0$, then $\int_1^\infty t^{x-1}e^{-t}\,dt$ is improper convergent of the first kind.[25]

Solution. Since

$$\lim_{t\to\infty} \frac{t^{x+1}}{e^t} = 0,$$

there is some $c \geq 1$ so that

$$\frac{t^{x+1}}{e^t} \leq 1$$

for all $t \geq c$. From this it follows that $e^{-t}t^{x-1} \leq t^{-2}$ for $t \geq c$. Since we know that $\int_c^\infty t^{-2}\,dt$ converges by Exercise 2, $\int_c^\infty e^{-t}t^{x-1}\,dt$ also converges by comparison test 1. The fact that $e^{-t}t^{x-1}$ is integrable on $[1, c]$ implies that

$$\int_1^\infty t^{x-1}e^{-t}\,dt = \int_1^c t^{x-1}e^{-t}\,dt + \int_c^\infty t^{x-1}e^{-t}\,dt$$

and therefore $\int_1^\infty t^{x-1}e^{-t}\,dt$ converges. ■

This completes the proof that the integral $\int_0^\infty t^{x-1}e^{-t}\,dt$ converges for all $x > 0$. We next consider how to find values of this integral for given values of x.

Example 3. $\Gamma(1) = \int_0^\infty e^{-t}\,dt = \lim_{k\to\infty} -\frac{1}{e^t}\Big|_0^k = \lim_{k\to\infty}\left(-\frac{1}{e^k} + 1\right) = 1.$

In Exercise 3, you are asked to use integration by parts to show that $\Gamma(2) = 1$. In a similar way, we can show that $\Gamma(3) = 2$ and $\Gamma(4) = 6 = 3!$. These results suggest the conjecture that $\Gamma(n) = (n-1)!$. A proof of this result follows from the recursion formula obtained in Example 4.

Example 4. If $x > 0$, then $\Gamma(x+1) = x\,\Gamma(x)$.

Solution. Using integration by parts, we have

$$\Gamma(x+1) = \int_0^\infty t^x e^{-t}\,dt$$

---

[25] If $f(x)$ is integrable on intervals of the form $[a, t]$ for all $t > a$, then $\int_a^\infty f(x)\,dx$ is an *improper integral of the first kind* and we define $\int_a^\infty f(x)\,dx = \lim_{t\to\infty}\int_a^t f(x)\,dx$.

$$
\begin{aligned}
&= \lim_{k\to\infty} \int_0^k t^x e^{-t}\,dt \\
&= \lim_{k\to\infty} \left.\frac{-t^x}{e^t}\right|_0^k + \lim_{k\to\infty} \int_0^k x t^{x-1} e^{-t}\,dt \\
&= \lim_{k\to\infty} \left(\frac{-k^x}{e^k}\right) + x \int_0^\infty t^{x-1} e^{-t}\,dt \\
&= 0 + x\,\Gamma(x).
\end{aligned}
$$

■

From this recursion result, it follows that $\Gamma(n) = (n-1)!$ for $n \in \mathbf{N}$ (see Exercise 5). This means the gamma function is a generalization of the factorial function, which is defined only for natural numbers. A next step is to find a value for $\Gamma(\frac{1}{2})$.

Example 5.

$$
\begin{aligned}
\Gamma\left(\frac{1}{2}\right) &= \int_0^\infty t^{-1/2} e^{-t}\,dt \\
&= \int_0^\infty \frac{1}{x} e^{-x^2} 2x\,dx \quad \text{by the substitution } t = x^2 \\
&= 2\int_0^\infty e^{-x^2}\,dx \\
&= \sqrt{\pi} \quad \text{by calculus review problem 20 in Part I.}
\end{aligned}
$$

Using the recursion formula in Example 4, we also know that

$$
\begin{aligned}
\Gamma\left(\frac{3}{2}\right) &= \frac{1}{2}\sqrt{\pi}, \\
\Gamma\left(\frac{5}{2}\right) &= \frac{3}{2}\Gamma\left(\frac{3}{2}\right) = \frac{3}{4}\sqrt{\pi},
\end{aligned}
$$

and in general,

$$
\Gamma\left(2k + \frac{1}{2}\right) = \frac{1 \cdot 3 \cdots (2k-1)}{2^k}\sqrt{\pi}.
$$

The value for a function defined by an integral can be represented geometrically as the area of a certain region. The value for

$$
\ln x = \int_1^x \frac{1}{t}\,dt
$$

is represented by the area of the region under the graph of $f(t) = 1/t$ and over the interval $[1, x]$ when $x > 1$. For different values of x, the function remains the same, while the base of the region changes. The value for the gamma function $\Gamma(x) = \int_0^\infty t^{x-1}e^{-t}\,dt$ is represented by the area of the unbounded region in the first quadrant that is under the graph of $f(t) = t^{x-1}e^{-t}$. For different values of x, the function changes, while the interval $[0, \infty)$ remains the same.

Example 6. For natural numbers n, the value of $n!$ for $\Gamma(n+1)$ is represented by the area of the unbounded region in the first quadrant that is under the graph of $f_n(t) = t^n e^{-t}$. This function has a relative maximum at the point $(n, (n/e)^n)$ and inflection points at $t = n \pm \sqrt{n}$, when $n > 1$.

Solution.

$$f_n'(t) = t^n e^{-t}(-1) + nt^{n-1}e^{-t} = t^{n-1}e^{-t}(n-t) = 0$$

implies that $f_n(t)$ has a relative maximum at

$$\left(n, \left(\frac{n}{e}\right)^n\right),$$

since

$$f_n(n) = n^n e^{-n} = \left(\frac{n}{e}\right)^n.$$

In a similar way, we can show that $f_n(t)$ has points of inflection at $t = n \pm \sqrt{n}$ when $n > 1$ (see Exercise 7).

To illustrate, $f_4(t) = t^4 e^{-t}$ has a relative maximum at

$$\left(4, \left(\frac{4}{e}\right)^4\right) \approx (4, 4.68)$$

with points of inflection at $(2, 2^4 e^{-2}) \approx (2, 2.165)$ and at $(6, 6^4 e^{-6}) \approx (6, 3.212)$ (see Figure 6). ■

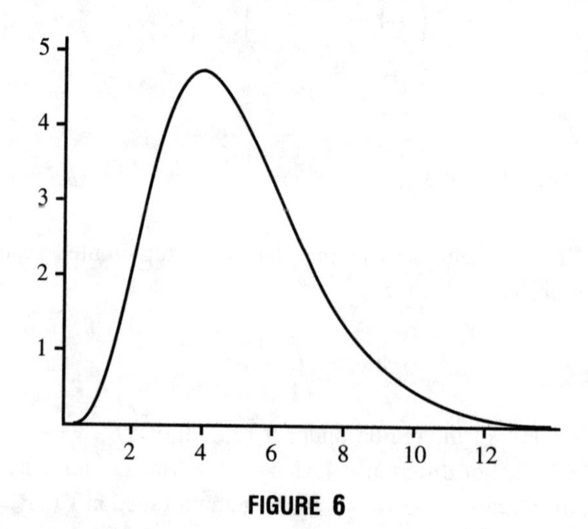

FIGURE 6

The beta function is another special function that can be defined in terms of an integral by the formula

$$\beta(x, y) = \int_0^1 t^{x-1}(1-t)^{y-1}\, dt.$$

One of the more interesting results[26] among the special functions is that

$$\beta(x, y) = \frac{\Gamma(x)\Gamma(y)}{\Gamma(x+y)}.$$

Example 7. Find the area in the first quadrant under the graph of $\sqrt{x} + \sqrt{y} = 1$.

Solution. By using calculus, we can find this area by

$$\begin{aligned}\int_0^1 (1-\sqrt{x})^2\, dx &= \int_0^1 (1 - 2x^{1/2} + x)\, dx \\ &= \left(x - \frac{4}{3}x^{2/3} + \frac{1}{2}x^2\right)\Bigg|_0^1 = \frac{1}{6}.\end{aligned}$$

We can also solve this problem by use of the beta function as follows.

$$\begin{aligned}\int_0^1 (1-\sqrt{x})^2\, dx &= \int_0^1 (1-u)^2\, 2u\, du \quad \text{by use of the substitution } u = \sqrt{x} \\ &= 2\,\beta(2,3) \\ &= \frac{2\,\Gamma(2)\Gamma(3)}{\Gamma(5)} \\ &= \frac{2 \cdot 1 \cdot 2}{24} = \frac{1}{6}\end{aligned}$$

■

Example 8. Find the area in the first quadrant under the graph of $\sqrt{x} + \sqrt{y} = \sqrt{a}$.

Solution.

$$\begin{aligned}\text{Area} = \int_0^a (\sqrt{a} - \sqrt{x})^2\, dx &= \int_0^a a\left(1 - \sqrt{\frac{x}{a}}\right)^2 dx \\ &= a^2 \int_0^1 2u(1-u)^2\, du \quad \text{by the substitution } u = \sqrt{\frac{x}{a}} \\ &= a^2 \cdot \frac{1}{6}\end{aligned}$$

by the result in Example 7. ■

---

[26] For a proof of this result, see R. Creighton Buck, *Advanced Calculus*, McGraw-Hill Book Company, New York, 1965, pp. 218–9.

Example 9. Several formulas are given below to show the diversity of integrals that can be evaluated by use of the gamma and beta functions.

$$\int_{-\infty}^{\infty} e^{-x^n}\,dx = \frac{2}{n}\Gamma\left(\frac{1}{n}\right) \tag{1}$$

$$\int_{0}^{\infty} x^p e^{-x^q}\,dx = \frac{1}{q}\Gamma\left(\frac{p+q-1}{q}\right) \quad \text{when } q>0 \text{ and } p+q>1 \tag{2}$$

$$\int_{-\infty}^{\infty} \frac{e^{at}}{1+e^t}\,dt = \Gamma(a)\,\Gamma(1-a) \quad \text{by using the substitution } u=\frac{1}{1+e^t} \tag{3}$$

$$\int_{0}^{1} \frac{dx}{\sqrt{1-x^4}} = \frac{1}{4}\beta\left(\frac{1}{4},\frac{1}{2}\right) \tag{4}$$

$$\int_{0}^{\pi/2} \sqrt{\sin\theta}\,d\theta = \frac{1}{2}\beta\left(\frac{3}{4},\frac{1}{2}\right) \tag{5}$$

$$\int_{0}^{\infty} u^p e^{-u^q}\,du = \frac{1}{q}\Gamma\left(\frac{p+1}{q}\right) \tag{6}$$

$$\int_{0}^{\infty} \frac{e^{-2x}}{\sqrt{x}}\,dx = \sqrt{\frac{\pi}{2}} \tag{7}$$

In order to obtain answers for many of the integrals in Example 9, we need to approximate values of $\Gamma(r)$, where r is a rational number. The usual method for this is to use a power series expansion of $\Gamma(x)$ or $\ln\Gamma(x+1)$.

Exercises

1. Prove that

$$\int_0^1 \frac{1}{t^p}\,dt \quad \text{converges to} \quad \frac{1}{1-p} \quad \text{if } 0<p<1.$$

2. Prove that

$$\int_a^{\infty} \frac{1}{t^p}\,dt \quad \text{converges to} \quad \frac{1}{(p-1)a^{p-1}} \quad \text{if} \quad a>0 \quad \text{and} \quad p>1.$$

3. Prove Comparison Test 1.

4. Use integration by parts to verify that

$$\Gamma(2) = \int_0^{\infty} te^{-t}\,dt = 1.$$

5. Use mathematical induction to prove $\Gamma(n)=(n-1)!$ for positive integers n.

6. Verify that

$$\Gamma\left(\frac{1}{3}\right) = 3\int_0^{\infty} e^{-x^3}\,dx.$$

7. Verify that $f_n(t) = t^n e^{-t}$ has points of inflection at $t = n \pm \sqrt{n}$, when $n > 1$. See Example 6.

8. Show that the area in the first quadrant under the curve

$$\sqrt[3]{x} + \sqrt[3]{y} = \sqrt[3]{a} \quad \text{is} \quad \frac{a^2}{20}.$$

Use the work in Example 8 as a guide.

9. Prove that the area in the first quadrant under the curve

$$\sqrt[n]{x} + \sqrt[n]{y} = \sqrt[n]{a} \quad \text{is} \quad \frac{(a\, n!)^2}{(2n)!}.$$

10. Prove some of the integral formulas listed in Example 9.

Problem 30. Bernoulli Numbers

The Bernoulli numbers are an unusual sequence of rational numbers that begin with the following terms. $B_n = 0$ for all odd $n \geq 3$ and

$$\begin{aligned}
B_0 &= 1 \\
B_1 &= -\frac{1}{2} \\
B_2 &= \frac{1}{6} \\
B_4 &= -\frac{1}{30} \\
B_6 &= \frac{1}{42} \\
B_8 &= -\frac{1}{30} \\
B_{10} &= \frac{5}{66} \\
B_{12} &= -\frac{691}{2730} \\
B_{14} &= \frac{7}{6} \\
B_{16} &= -\frac{3617}{510} \\
B_{18} &= \frac{43867}{798} \\
B_{20} &= -\frac{174611}{330}
\end{aligned}$$

Though we might think we see some pattern in the early terms, this is rudely interrupted by B_{12} and B_{16}, and then totally destroyed by the following term.

$$B_{60} = -\frac{1215233140483755572040304994079820246041491}{56786730}$$

No matter how much explanation is given, there will always be an aura of mystery surrounding the Bernoulli numbers, and why they should occur in so many diverse settings. They were named for James Bernoulli after he presented these numbers in his book *Ars conjectandi* (The Art of Conjecturing) which was published in 1713. Bernoulli acknowledged that he learned of these numbers through the book *Academiae Algebrae* published by Johann Faulhaber in 1631.[27] Faulhaber introduced these numbers in connection with his investigation of a formula for the finite sum $\sum_{i=1}^{n} i^k$ for natural numbers k. Note these are the same formulas needed to find the area under the curve $y = x^k$ by the method identified as Cavalieri sums in Problem 4.

Example 1. The usual method for defining Bernoulli numbers is to use the Maclaurin series expansion for the function $x/(e^x - 1)$, and identify

$$\frac{x}{e^x - 1} = \sum_{n=0}^{\infty} \frac{B_n}{n!} x^n.$$

Since

$$\begin{aligned}\frac{x}{e^x - 1} &= \frac{x}{(1 + x + \frac{x^2}{2!} + \frac{x^3}{3!} + \cdots) - 1} \\ &= \frac{1}{1 + \frac{x}{2!} + \frac{x^2}{3!} + \cdots} \\ &= 1 - \frac{x}{2} + \frac{x^2}{12} - \frac{x^4}{720} + \cdots\end{aligned}$$

it follows that

$$B_0 = 1, \quad B_1 = -\frac{1}{2}, \quad \frac{B_2}{2!} = \frac{1}{12}, \quad \frac{B_3}{3!} = 0, \quad \text{and} \quad \frac{B_4}{4!} = -\frac{1}{720}$$

so that $B_0 = 1$, $B_1 = -\frac{1}{2}$, $B_2 = \frac{1}{6}$, $B_3 = 0$, and $B_4 = -\frac{1}{30}$.

This process of long division, which is also used in calculus to find the first terms of the Maclaurin series for $\tan x$ (see Problem 3) is very tedious. A better method is to use the recursion formula developed in the reading of Selection 5, namely that

$$\sum_{k=0}^{n-1} \binom{n}{k} B_k = \binom{n}{0} B_0 + \binom{n}{1} B_1 + \binom{n}{2} B_2 + \cdots + \binom{n}{n-1} B_{n-1} = 0.$$

[27] See pp. 106–9 in [8].

Example 2. If we have already obtained the values for B_k when $0 \le k \le 3$, then we have for $n = 5$,

$$\begin{aligned}\tbinom{5}{0} B_0 + \tbinom{5}{1} B_1 + \tbinom{5}{2} B_2 + \tbinom{5}{3} B_3 + \tbinom{5}{4} B_4 &= 0\\ B_0 + 5\,B_1 + 10\,B_2 + 10\,B_3 + 5\,B_4 &= 0\\ 1 - \tfrac{5}{2} + \tfrac{5}{3} + 0 + 5\,B_4 &= 0\end{aligned}$$

from which it follows that $B_4 = -\frac{1}{30}$.

We next list some of the general formulas that can be expressed in terms of the Bernoulli numbers.

$$\sum_{n=1}^{\infty} \frac{1}{n^{2k}} = \frac{(2\pi)^{2k} |B_{2k}|}{2(2k)!} \qquad \text{(see Selections 2 and 5)} \qquad (1)$$

$$\tan x = \sum_{k=1}^{\infty} (-1)^{k+1} \frac{2^{2k}(2^{2k}-1)\,B_{2k}}{(2k)!} x^{2k-1} \qquad \text{(see Selection 5)} \qquad (2)$$

$$\sum_{i=1}^{n} i^k = \frac{1}{k+1} \sum_{j=0}^{k} \binom{k+1}{j} B_j\, n^{k+1-j} \qquad \text{(see footnote 27)} \qquad (3)$$

$$\sum_{i=1}^{n} f(i) = \left(\int f(x)\,dx + \frac{1}{2} f(x) + \sum_{n=1}^{\infty} \frac{B_{2n}}{(2n)!} f^{(2n-1)}(x) \right)_{x=n} \qquad (4)$$

Refer to Selection 2 or to **[23]** for further information about the Euler–Maclaurin formula that is expressed in equation (4) above.

Exercises

1. Use the method of long division to find the first four Bernoulli numbers as presented in Example 1.

2. Use the recursion formula presented in Example 2 to obtain the values for B_6, B_8, and B_{10}.

3. Use the general formula presented for $\tan x$ in equation (2) to verify that the first four terms of the Maclaurin series for $\tan x$ are

$$x + \frac{1}{3}x^3 + \frac{2}{15}x^5 + \frac{17}{315}x^7.$$

4. Use the general formula presented in equation (1) to obtain the formulas

$$\sum_{n=1}^{\infty} \frac{1}{n^2} = \frac{\pi^2}{6}, \quad \sum_{n=1}^{\infty} \frac{1}{n^4} = \frac{\pi^4}{90}, \quad \text{and} \quad \sum_{n=1}^{\infty} \frac{1}{n^6} = \frac{\pi^6}{945}.$$

5. Use the general formula in equation (3) to obtain the formula

$$\sum_{i=1}^{n} i^4 = \frac{1}{30} n(n+1)(2n+1)(3n^2+3n-1)$$

which we obtained in Problem 4 by another method.

Problem 31. A Look at Fourier Series

Maclaurin series and Fourier series are the most familiar examples of a representation of a function by an infinite series. The development of Maclaurin series (or power series) occurred about a century before the development of Fourier series (or trigonometric series). Brook Taylor presented the concept of a Taylor series in his 1715 book titled *Methodus Incrementorum Directa et Inversa*, while Joseph Fourier presented his series in an 1822 publication. Both series were explored well before these formal presentations, as Newton used power series in the 1660s and Euler used trigonometric series in the mid-1700s.

Power series provide a way to express standard continuous functions from calculus as the limit of a sequence of polynomials. These series are used to approximate values such as $\ln 2$, $\sqrt{3}$, e, π, and $\int_0^1 e^{x^2}\,dx$, and have been used to establish other results such as

$$\sum_{n=1}^{\infty} \frac{1}{n^2} = \frac{\pi^2}{6}$$

and that e is irrational (see Problems 11 and 23).

Trigonometric series allow us to represent functions with discontinuities, and are useful in the study of topics such as the vibrating string or fluid mechanics. While Georg Cantor was investigating some convergence properties of Fourier series, he was led to his initial consideration of infinite sets and cardinality.

Example 1. We begin our derivation of the usual formulas for the coefficients of a Fourier series by assuming that an integrable function $f(x)$ can be represented over the interval $[-\pi, \pi]$ by a trigonometric series of the form

$$f(x) = \sum_{n=1}^{\infty} A_n \sin nx + \sum_{n=0}^{\infty} B_n \cos nx.$$

The following steps are then carried out to determine the formula for A_k, assuming such operations as interchanging the order of integration and summation are valid. This is true whenever there is uniform convergence for the series.

$$\begin{aligned} f(x)\sin kx &= \sum_{n=1}^{\infty} A_n \sin nx \sin kx + \sum_{n=0}^{\infty} B_n \cos nx \sin kx \\ \int_{-\pi}^{\pi} f(x)\sin kx\,dx &= \sum_{n=1}^{\infty} \int_{-\pi}^{\pi} A_n \sin nx \sin kx\,dx + \sum_{n=0}^{\infty} \int_{-\pi}^{\pi} B_n \cos nx \sin kx\,dx \\ &= A_k \cdot \pi \end{aligned}$$

by use of formulas from Problem 1, which show that all but one of the integrals in the summations above are equal to 0. When this is combined with Exercise 1, we have

$$A_k = \frac{1}{\pi}\int_{-\pi}^{\pi} f(x)\sin kx\,dx$$

$$B_k = \frac{1}{\pi}\int_{-\pi}^{\pi} f(x)\cos kx\,dx \qquad \text{if } k \geq 1$$

$$B_0 = \frac{1}{2\pi}\int_{-\pi}^{\pi} f(x)\,dx.$$

Observe that the coefficients of the Maclaurin series representation of $f(x)$ are represented in terms of the derivatives of $f(x)$, while the coefficents of the Fourier series representation of $f(x)$ are represented in terms of an integral involving $f(x)$. Another item of interest is that the power series expansions can be viewed as elements of a vector space over the reals having the set $\{1, x, x^2, \ldots\}$ as its basis, while the Fourier series expansions can be considered as elements of a vector space over the reals having the set $\{1, \sin x, \cos x, \sin 2x, \cos 2x, \ldots\}$ as its basis.

Example 2. In order to find the Fourier series representation for $f(x) = |x|$, we observe that all $A_n = 0$ since $|x|$ is an even function, while

$$B_0 = \frac{1}{2\pi}\int_{-\pi}^{\pi} |x|\,dx = \frac{\pi}{2}$$

and for $n \geq 1$,

$$\begin{aligned}B_n &= \frac{1}{\pi}\int_{-\pi}^{\pi} |x|\cos nx\,dx = \frac{2}{\pi}\int_{0}^{\pi} x\cos nx\,dx\\ &= \frac{2}{\pi}\left(\frac{x\sin nx}{n} + \frac{\cos nx}{n^2}\right)\Bigg|_0^{\pi} = \frac{2}{\pi}\left(\frac{\cos n\pi}{n^2} - \frac{\cos 0}{n^2}\right)\\ &= \frac{2}{\pi}\frac{(-1)^n - 1}{n^2},\end{aligned}$$

so the desired Fourier series expansion is

$$\begin{aligned}|x| &= \frac{\pi}{2} + \sum_{n=1}^{\infty} \frac{2}{\pi}\frac{(-1)^n - 1}{n^2}\cos nx\\ &= \frac{\pi}{2} - \frac{4}{\pi}\left(\cos x + \frac{\cos 3x}{3^2} + \frac{\cos 5x}{5^2} + \cdots\right).\end{aligned}$$

Example 3. In order to find the Fourier series representation for $f(x) = x^3$, we observe that all $B_n = 0$ since x^3 is an odd function, while

$$A_n = \frac{1}{\pi}\int_{-\pi}^{\pi} x^3\sin nx\,dx = \frac{2}{\pi}\int_{0}^{\pi} x^3\sin nx\,dx$$

$$= \frac{2}{\pi}\left(\frac{-x^3\cos nx}{n} + \frac{3x^2\sin nx}{n^2} + \frac{6x\cos nx}{n^3} - \frac{6\sin nx}{n^4}\right)\Bigg|_0^{\pi}$$

$$= \frac{2}{\pi}\left(\frac{-\pi^3\cos n\pi}{n} + \frac{6\pi\cos n\pi}{n^3}\right)$$

$$= \frac{(-1)^{n+1}2\pi^2}{n} + \frac{(-1)^n 12}{n^3}$$

and so the desired Fourier series expansion is

$$x^3 = \sum_{n=1}^{\infty}\left(\frac{(-1)^{n+1}2\pi^2}{n} + \frac{(-1)^n 12}{n^3}\right)\sin nx.$$

When $x = \pi/2$ is substituted into this formula, we obtain

$$\frac{\pi^3}{8} = 2\pi^2\sum_{n=1}^{\infty}\frac{(-1)^{n+1}\sin\frac{n\pi}{2}}{n} + 12\sum_{n=1}^{\infty}\frac{(-1)^n\sin\frac{n\pi}{2}}{n^3}$$

$$= 2\pi^2\left(1 - \frac{1}{3} + \frac{1}{5} - \frac{1}{7} + - \cdots\right) + 12\left(-1 + \frac{1}{3^3} - \frac{1}{5^3} + \frac{1}{7^3} - + \cdots\right)$$

$$= 2\pi^2\left(\frac{\pi}{4}\right) - 12\left(1 - \frac{1}{3^3} + \frac{1}{5^3} - \frac{1}{7^3}\cdots\right)$$

from which it follows that

$$1 - \frac{1}{3^3} + \frac{1}{5^3} - \frac{1}{7^3} + \cdots = \frac{\pi^3}{32}.$$

Example 4. The standard way to find the Fourier series for $f(x) = \sin^2 x$ is to evaluate the coefficients

$$B_n = \frac{1}{\pi}\int_{-\pi}^{\pi}\sin^2 x\cos nx\,dx.$$

We know that

$$A_n = \frac{1}{\pi}\int_{-\pi}^{\pi}\sin^2 x\sin nx\,dx = 0$$

because $\sin^2 x \sin nx$ is an odd function. It is fairly routine to use antiderivatives to show that

$$B_0 = \frac{1}{2\pi}\int_{-\pi}^{\pi}\sin^2 x\,dx = \frac{1}{2}$$

$$B_1 = \frac{1}{\pi}\int_{-\pi}^{\pi}\sin^2 x\cos x\,dx = 0$$

$$B_2 = \frac{1}{\pi}\int_{-\pi}^{\pi}\sin^2 x\cos 2x\,dx = -\frac{1}{2}.$$

But since we have the trigonometric identity $\sin^2 x = \frac{1}{2} - \frac{1}{2}\cos 2x$, it follows that $B_n = 0$ for all $n \geq 3$ without the need for additional integration.

Exercises

1. Establish the general formulas given above for B_0 and B_k for $k \geq 1$ by adjusting the work given in Example 1 for derivation of the formula for A_k.

2. Show how to verify that the Fourier series expansion for $f(x) = x^2$ is given by

$$x^2 = \frac{\pi^2}{3} + \sum_{n=1}^{\infty} \frac{(-1)^n 4}{n^2} \cos nx.$$

3. Substitute $x = \pi$ into the Fourier series expansion for $f(x) = x^2$ given in Exercise 2 in order to obtain the well-known formula that

$$\sum_{n=1}^{\infty} \frac{1}{n^2} = \frac{\pi^2}{6}.$$

4. The result in Exercise 3 gives hope that the Fourier series expansion for $f(x) = x^3$ might be used to find a formula for

$$\sum_{n=1}^{\infty} \frac{1}{n^3}.$$

By looking at the work in Example 3, explain why this does not happen. This problem remains one of the famous problems in analysis that is still unsolved.

5. Verify the values for the three integrals in Example 4 by use of antiderivatives.

6. **(a)** Obtain the trigonometric identity $\cos 3x = 4\cos^3 x - 3\cos x$ and use it to obtain the Fourier series for $\cos^3 x$ in a manner similar to that shown in Example 4.

   (b) Use the Fourier series $\cos^3 x = \frac{3}{4}\cos x + \frac{1}{4}\cos 3x$ in order to find the values of the following four integrals without using methods of integration.

$$\int_{-\pi}^{\pi} \cos^3 x\, dx =$$

$$\int_{-\pi}^{\pi} \cos^4 x\, dx =$$

$$\int_{-\pi}^{\pi} \cos^3 x \cos 2x\, dx =$$

$$\int_{-\pi}^{\pi} \cos^3 x \cos 3x\, dx =$$

Problem 32. Summability of Divergent Series

Calculus students are quick to learn that the word *divergent* carries bad connotations. However, this view is not universally held among mathematicians. The British mathematician

G. H. Hardy (see Essay 6) wrote an entire book on the subject of divergent series, and he regarded this as part of his best work. There are divergent series for which the early terms of the sequence of partial sums approximate some physical phenomena quite accurately before the divergence of later terms appears. This problem introduces the idea that some divergent series possess a property of *summability*, if the arithmetic mean sequence of its sequence of partial sums converges.

A similar situation occurs in integration. We may have a function that is not Riemann-integrable, but may be integrable under a different definition, such as the one by Lebesgue. Or integrals such as

$$\int_0^1 \frac{1}{\sqrt{x}}\,dx \quad \text{or} \quad \int_1^\infty \frac{1}{x^2}\,dx$$

may not exist in the Riemann sense, but are improper integrals that converge to a finite value.

Definition 1. The series $\sum_{n=1}^\infty a_n$ is *divergent* if $\{s_n\}$ where $s_n = a_1 + \cdots + a_n$ is a divergent sequence. If the sequence $\{A_n\}$, where

$$A_n = \frac{1}{n}\sum_{i=1}^n s_i$$

converges to A, then the series $\sum_{n=1}^\infty a_n$ is said to be *summable* to A.

Example 1. One example that puzzled 18th century mathematicians was the series

$$\sum_{n=1}^\infty (-1)^{n+1} = 1 - 1 + 1 - 1 + - \cdots.$$

Arguments were made by grouping of terms as shown below that its value should be 0 or 1 or $\frac{1}{2}$.

$$\sum_{n=1}^\infty (-1)^{n+1} = (1-1) + (1-1) + (1-1) + \cdots = 0$$

$$\sum_{n=1}^\infty (-1)^{n+1} = 1 - (1-1) - (1-1) - \cdots = 1$$

$$\sum_{n=1}^\infty (-1)^{n+1} = 1 - \sum_{n=1}^\infty (-1)^{n+1}$$

A second reason given for the value of $\frac{1}{2}$ is that it is the average of the values of 0 and 1. A third reason is to appeal to the series expansion

$$\frac{1}{1+x} = 1 - x + x^2 - x^3 + \cdots$$

when $x = 1$.

A modern answer, based on the definition of series convergence, is to simply say that the series $\sum_{n=1}^{\infty}(-1)^{n+1}$ diverges because the sequence of partial sums $\{s_n\}$ diverges by oscillation. But the sequence $\{A_n\}$ of arithmetic means for the sequence of partial sums, namely

$$A_n = \frac{1}{n}\sum_{i=1}^{n} s_i,$$

clearly converges to $\frac{1}{2}$, and so the series is summable to $\frac{1}{2}$. Observe the following details which support this conclusion.

$$\sum_{n=1}^{\infty} a_n = 1 - 1 + 1 - 1 + 1 - 1 + - \cdots$$

$$\{s_n\} = 1,\ 0,\ 1,\ 0,\ 1,\ 0,\ \cdots$$

$$\{A_n\} = 1,\ \frac{1}{2},\ \frac{2}{3},\ \frac{2}{4},\ \frac{3}{5},\ \frac{3}{6}, \cdots \quad \text{so } A_{2n} = \frac{1}{2} \quad \text{and} \quad A_{2n-1} = \frac{n}{2n-1}$$

Note that the series whose sequence of partial sums is $\{A_n\}$ is

$$\sum_{n=1}^{\infty} b_n = 1 - \frac{1}{2} + \frac{1}{6} - \frac{1}{6} + \frac{1}{10} - \frac{1}{10} + - \cdots.$$

Example 2. We begin with the power series

$$\frac{1+x}{1+x+x^2} = 1 - x^2 + x^3 - x^5 + x^6 - x^8 + x^9 - + \cdots,$$

which can be obtained by the following work.

$$\begin{aligned}
\frac{1+x}{1+x+x^2} &= \frac{1-x^2}{1-x^3} \qquad \text{if } x \neq 1 \\
&= (1-x^2)(1+x^3+x^6+x^9+\cdots) \qquad \text{if } -1 < x < 1 \\
&= 1 - x^2 + x^3 - x^5 + x^6 - x^8 + x^9 - + \cdots
\end{aligned}$$

Setting $x = 1$ yields $\frac{2}{3} = 1 - 1 + 1 - 1 + 1 - 1 + - \cdots$, adding more confusion to the material in Example 1. To see why the value of $\frac{2}{3}$ is reasonable, it is helpful to rewrite the series as

$$1 + 0\cdot x - x^2 + x^3 + 0\cdot x^4 - x^5 + x^6 + 0\cdot x^7 - x^8 + x^9 + \cdots.$$

Now when $x = 1$, we get the series $1 + 0 - 1 + 1 + 0 - 1 + 1 + 0 - 1 + \cdots$.

In Exercise 2, you are asked to prove that this series is summable to $\frac{2}{3}$. So by comparison with the series in Example 1, we see that the introduction of additional zeros into a series can affect the value.

Summability is an extension of the usual concept of series convergence, in that every convergent series is also summable, and to the same value. This result is stated below as Theorem 1.

Theorem 1. *Let the series $\sum_{n=1}^{\infty} a_n$ converge to a. Then $\sum_{n=1}^{\infty} a_n$ is also summable to a.*

Proof We let $s_n = a_1 + a_2 + \cdots + a_n$, and know that $\{s_n\}$ converges to a. We also set $S_n = s_1 + s_2 + \cdots + s_n$ and let

$$A_n = \frac{S_n}{n}.$$

We need to show that $\{A_n\}$ converges to a.

Case 1: $a = 0$

For $\epsilon > 0$, there is some N so that

$$|s_n| < \frac{\epsilon}{2} \quad \text{for } n \geq N.$$

For $n > N$,

$$\begin{aligned} |A_n| &= \left| \frac{s_1 + s_2 + \cdots + s_N + \cdots + s_n}{n} \right| \\ &\leq \left| \frac{s_1 + s_2 + \cdots + s_N}{n} \right| + \left| \frac{s_{N+1} + \cdots + s_n}{n} \right| \\ &\leq \frac{|S_N|}{n} + \frac{n-N}{n} \cdot \frac{\epsilon}{2}. \end{aligned}$$

So if

$$n > \frac{2|S_N|}{\epsilon},$$

then

$$|A_n| < \frac{|S_N|}{n} + \frac{\epsilon}{2} < \frac{\epsilon}{2} + \frac{\epsilon}{2} = \epsilon,$$

and the sequence $\{A_n\}$ converges to $a = 0$ as desired.

Case 2: $a \neq 0$

The proof for this case is left as an exercise. ■

It may happen that a divergent series is not summable, but repeating the process of forming the sequence of arithmetic means leads to a convergent sequence. This is illustrated in Example 3. In this case, we say the original series is C_2-summable, and let C_1-summable mean the same thing as summable. The concept can be extended to C_k-summable for any natural number k, depending on how many times the process of forming the sequence of arithmetic means needs to be repeated. This notation reflects the work of Ernesto Cesàro (1859–1906), an Italian mathematician who worked on this problem early in his career. Other mathematicians developed alternate definitions of summability, but we will not consider them in this problem.[28] We state the following theorem, leaving its proof for an exercise, and use it in Example 3.

Theorem 2. *Let the sequence* $\{A_n\}$ *be such that* $\{A_{2n}\}$ *converges to* a *and* $\{A_{2n-1}\}$ *converges to* b. *Then the sequence* $\{A'_n\}$ *where*

$$A'_n = \frac{1}{n}\sum_{i=1}^{n} A_i$$

converges to $(a+b)/2$.

Proof This is left for an exercise. ■

Notice that Theorem 1 is a special case of Theorem 2. A proof for Theorem 2 can be obtained by adjustments to the proof given for Theorem 1. Generalizations of Theorem 2 are possible if there are more than two cluster points of $\{A_n\}$.

Example 3. Let $\sum_{n=1}^{\infty} a_n = \sum_{n=1}^{\infty}(-1)^{n+1}n = 1-2+3-4+5-6+-\cdots$.

$$\{s_n\} = 1, -1, 2, -2, 3, -3, \cdots$$

$$\{A_n\} = 1, 0, \tfrac{2}{3}, 0, \tfrac{3}{5}, 0, \tfrac{4}{7}, 0, \cdots$$

So the original series $\sum_{n=1}^{\infty} a_n$ is not C_1-summable, but it is C_2-summable to

$$\frac{0+\frac{1}{2}}{2} = \frac{1}{4}$$

by Theorem 2.

If Theorem 2 were not available, we would have to form the sequence of arithmetic means $\{A'_n\}$ for the sequence $\{A_n\}$. By looking at the beginning terms below, we see this sequence is decreasing, but it is not possible to easily determine the value to which it converges.

$$\{A'_n\} = 1, \frac{1}{2}, \frac{5}{9}, \frac{5}{12}, \frac{34}{75}, \frac{34}{90}, \frac{298}{735}, \frac{298}{840}, \cdots$$

Example 4. The harmonic series

$$\sum_{n=1}^{\infty} \frac{1}{n}$$

is not summable.

---

[28]For additional information on summability, see David V. Widder, *Advanced Calculus*, 2nd ed., Prentice-Hall, Inc., Englewood Cliffs, 1961, pp. 307–313.

Solution. Since the sequence $\{s_n\}$ of partial sums is $1, 1+\frac{1}{2}, 1+\frac{1}{2}+\frac{1}{3}, \ldots,$ we obtain the following.

$$\begin{aligned} A_n &= \frac{1}{n}\left(1+\left(1+\frac{1}{2}\right)+\left(1+\frac{1}{2}+\frac{1}{3}\right)+\cdots+\left(1+\frac{1}{2}+\frac{1}{3}+\cdots+\frac{1}{n}\right)\right) \\ &= \frac{n}{n}\cdot 1+\frac{n-1}{n}\cdot\frac{1}{2}+\frac{n-2}{n}\cdot\frac{1}{3}+\cdots+\frac{2}{n}\cdot\frac{1}{n-1}+\frac{1}{n}\cdot\frac{1}{n} \\ &= 1+\left(1-\frac{1}{n}\right)\frac{1}{2}+\left(1-\frac{2}{n}\right)\frac{1}{3}+\cdots+\left(1-\frac{n-2}{n}\right)\frac{1}{n-1}+\left(1-\frac{n-1}{n}\right)\frac{1}{n} \\ &= \left(1+\frac{1}{2}+\frac{1}{3}+\cdots+\frac{1}{n}\right)-\left(\frac{1}{n}\left(\frac{1}{2}+\frac{2}{3}+\cdots+\frac{n-1}{n}\right)\right) \end{aligned}$$

We next observe that $\{b_n\}$ where

$$b_n = \frac{1}{n}\left(\frac{1}{2}+\frac{2}{3}+\cdots+\frac{n-1}{n}\right)$$

is bounded between 0 and 1. Since $A_n = s_n - b_n$ and since $\{s_n\}$ diverges to ∞, it follows that $\{A_n\}$ also diverges to ∞, and so

$$\sum_{n=1}^{\infty}\frac{1}{n}$$

is not summable. ■

Exercises

1. Find the series whose sequence of partial sums begins with the terms

$$1, \frac{3}{2}, \frac{11}{6}, \frac{25}{12}, \frac{137}{60}, \frac{147}{60}, \ldots.$$

2. In Example 2, verify that the series $1+0-1+1+0-1+1+0-1+\cdots$ is summable to $\frac{2}{3}$ by first finding expressions for A_{3n}, A_{3n-1}, and A_{3n-2}.

3. Show that the series

$$\sum_{n=1}^{\infty}\sin\frac{n\pi}{2}$$

is summable to $\frac{1}{2}$.

4. Verify the beginning terms of the sequence $\{A'_n\}$ in Example 3.

5. For the sequence $\{b_n\}$ in Example 4, prove that $0 < b_n < 1$ for all n.

6. Verify that the series

$$\sum_{n=1}^{\infty} a_n = 1 - 1 + 2 - 2 + 3 - 3 + 4 - 4 + \cdots$$

is not summable because $\{A_n\}$ diverges to infinity. Is it C_2-summable?

7. For Theorem 1, apply the result proven in Case 1 to the sequence $\{s'_n\}$ where $s'_n = s_n - a$ in order to obtain a proof for Case 2.

8. Prove Theorem 2.

Problem 33. Lebesgue Measure

A frequently used technique in analysis is to define a quantity as the infimum of all its upper bounds, or as the supremum of all its lower bounds. This is how the Riemann integral is usually defined. We start with an arbitrary partition P of the interval $[a, b]$ and use $U(f, P)$ to represent the upper sum $\sum_{i=1}^{n} M_i \Delta x_i$ and $L(f, P)$ to represent the lower sum $\sum_{i=1}^{n} m_i \Delta x_i$. We define $M_i = \sup\{f(x) : x \in [x_{i-1}, x_i]\}$ and $m_i = \inf\{f(x) : x \in [x_{i-1}, x_i)\}$, while $\Delta x_i = x_i - x_{i-1}$ is the length of the subinterval $[x_{i-1}, x_i]$. Then the upper integral

$$\overline{\int_a^b} f$$

is defined as the infimum of all the upper sum values $U(f, P)$, while the lower integral is defined as the supremum of all the lower sum values $L(f, P)$.

Early in the 20th century, Henri Lebesgue used this same technique to define the two concepts of the Lebesgue measure of a set and the Lebesgue integral of a function. In his definition of the Lebesgue integral, he replaced the expression $M_i \Delta x_i$ in Riemann's integral definition by $\sup\{f(x) : x \in A_i\} \cdot \mu(A_i)$ where the collection $\{A_i\}_{i=1}^{n}$ is a partition of $[a, b]$ into arbitrary sets and $\mu(A_i)$ is Lebesgue's measure of the set A_i, determined by generalizing the length of interval concept.

In a generalization, the measure μ should apply to more general sets than intervals I, and should assign the value $\mu(I) = b - a$ when the interval $I = (a, b)$ is measured. Here are some basic Axioms we would want such a measure to satisfy.

1. $\mu(S) \geq 0$
2. if $S \subset T$, then $\mu(S) \leq \mu(T)$
3. $\mu(S \cup T) = \mu(S) + \mu(T)$ if $S \cap T = \emptyset$
4. $\mu(S \cup T) \leq \mu(S) + \mu(T)$ for any sets S, T
5. $\mu(\cup_n S_n) \leq \sum_{n=1}^{\infty} \mu(S_n)$

We next introduce some terminology. For an arbitrary set S of real numbers, the sequence of open intervals $\{I_n\}$ where $I_n = (a_n, b_n)$ is called an *open cover* for S if $S \subset \cup_n I_n$ (recall the use of an open cover in the definition of a compact set). For any measure μ that satisfies the above five Axioms, we have

$$\mu(S) \leq \mu(\cup_{n=1}^{\infty} I_n) \leq \sum_{n=1}^{\infty} \mu(I_n) = \sum_{n=1}^{\infty} (b_n - a_n)$$

for every open cover $\{I_n\}$ of S, by the use of Axioms 2 and 5. Hence the value of the series $\sum_{n=1}^{\infty}(b_n - a_n)$ is an upper bound for $\mu(S)$.

After presenting the above notation, Lebesgue defined $\mu(S)$ in the following way.

$$\mu(S) = \inf\left\{\sum_{n=1}^{\infty}(b_n - a_n) : \{I_n = (a_n, b_n)\}_{n=1}^{\infty} \quad \text{is an open cover of } S\right\} \tag{1}$$

Example 1. The Lebesgue measure of any finite set is zero.

Solution. Let $S = \{a_1, a_2, \ldots, a_n\}$ and choose $\epsilon > 0$. Then the collection $\{I_i\}$ where

$$I_i = \left(a_i - \frac{\epsilon}{2n}, a_i + \frac{\epsilon}{2n}\right)$$

is an open cover for S. Also

$$\sum_{i=1}^{n} \ell I_i = \sum_{i=1}^{n} \frac{\epsilon}{n} = \epsilon.$$

Since $0 \leq \mu(S) < \epsilon$ holds for all $\epsilon > 0$, it follows that $\mu S = 0$. ■

For additional information, see Selection 4 in Part IV for a brief treatment of the life and work of Lebesgue. Consult [22] for an historical treatment of the development of the Lebesgue measure and integral, or one of the standard texts.[29]

Exercises

Use the definition of the Lebesgue measure $\mu(S)$ in (1) in order to prove the following results.

1. Explain why $\mu(S) = b - a$ if S is the open interval (a, b).
2. Prove that $\mu(S) = b - a$ if S is the closed interval $[a, b]$.
3. Prove that $\mu(S) = 0$ if S is a countably infinite set, such as the set of rational numbers.
   Hint: For any $\epsilon > 0$, find an open cover $\{I_n\}$ (where $I_n = (a_n, b_n)$) of the set

$$S = \{s_n : n \in \mathbf{N}\}$$

[29] For example, see Robert G. Bartle, *The Elements of Integration*, John Wiley & Sons, Inc., New York, 1966.

so that

$$\sum_{n=1}^{\infty} \ell(a_n, b_n) = \sum_{n=1}^{\infty} (b_n - a_n) < \epsilon.$$

4. Find $\mu(S)$ when $S = \{r : r \text{ is irrational and } 0 < r < 1\}$.

5. Prove that Axioms 1 and 3 imply Axiom 2. Also prove that Axioms 2 and 3 imply Axiom 4. This means Axioms 2 and 4 are dependent axioms, so they can be renamed as theorems.

Problem 34. Cantor's Middle Third Set

There are various ways to measure quantities in mathematics. In probability theory, we learn how to measure the probability of success of a given outcome. In calculus, we learn how to measure the area of a two-dimensional set by use of a definite integral. In the last third of the nineteenth century, two men came up with very different, yet important, ways to measure a set of real numbers. Georg Cantor's idea was to measure a set according to the number of elements that were in the set, which is called its cardinality. Henri Lebesgue measured a set by generalizing the concept of the length of an interval (see Problem 33).

Here are two examples. According to Cantor, the interval $[a, b]$ is uncountable and has cardinal number c. According to Lebesgue, the interval $[a, b]$ has length of $b - a$. For a second example, the set Q of rational numbers is a countable set with cardinal number $\aleph_o$ according to Cantor, while Lebesgue assigned it the measure of zero.

Early intuition suggested that a set is countable if and only if its Lebesgue measure is zero. Then Cantor came up with his example of an uncountable set that had Lebesgue measure zero. Supporting details are given below.

Start with $C_0 = [0, 1]$ and let C_1 be the set of points remaining when the open middle third of C_0 is deleted; i.e.,

$$C_1 = \left[0, \tfrac{1}{3}\right] \cup \left[\tfrac{2}{3}, 1\right].$$

Then let C_2 be the set obtained by deleting the middle third of each interval in C_1: i.e.,

$$C_2 = \left[0, \tfrac{1}{9}\right] \cup \left[\tfrac{2}{9}, \tfrac{1}{3}\right] \cup \left[\tfrac{2}{3}, \tfrac{7}{9}\right] \cup \left[\tfrac{8}{9}, 1\right].$$

Continue to define the sequence $\{C_n\}$ in this inductive way. Then $C = \cap_n C_n$ is called the Cantor set.

Exercises

1. Provide evidence that the Lebesgue measure $\mu(C)$ of the Cantor set C is 0.

2. Prove that C is an uncountable set. Hint: consider the ternary representation of the points in C.

Essays

Introduction. We present a historical overview for the development of calculus and analysis in six biographical essays, each based on a time period of this development. For each period, we choose the most significant individuals, and present details of their life and work. We briefly mention others who made a lesser contribution to analysis. Many outstanding mathematicians who did not contribute to calculus are omitted. These six essays are intentionally short, so that you will be encouraged to peruse all of them. Additional information can easily be found by looking at one of the standard histories of mathematics[1] or in the *Biographical Dictionary of Mathematics*. There are full biographies available for many mathematicians, although few of the main contributors to calculus have such a complete biography.

Most students of calculus believe it was Newton and Leibniz who discovered calculus, even if they are not quite certain what this entails. It is true that Newton and Leibniz saw patterns in the results obtained by their predecessors and had the insight that differentiation and integration—or finding slopes of tangent lines and finding areas under curves—were inverse processes. Essays 1 and 2 present these details. Essay 3 describes how their successors in the 18th century developed the results which constitute present-day calculus. Essays 4 and 5 describe attempts in the 19th century to provide a rigorous foundation for these results. In Essay 6, we present some 20th century developments which followed discoveries made at the end of the previous century, such as the infinite set theory of Cantor.

Essays 7 and 8 present an alternate perspective of two topics from calculus—the derivative formulas and the tests for series convergence. Students obtain the derivative formulas early in a calculus course before they are able to gain an understanding of their significance. In Essay 7, these formulas are obtained in a logical order, the way material is organized in an analysis course. We show three different ways to obtain the derivative for the natural logarithmic function. Infinite series constitute one of the most difficult parts of calculus for

[1] Some examples are [**6**], [**13**], and [**25**].

students to understand. The purpose of Essay 8 is to present proofs for the convergence tests in a way that shows their dependence upon each other and upon theorems involving sequences.

The final two essays are concerned with two of the most challenging parts of analysis, namely the nature of proof and the concept of topology. Essay 9 presents an informal discussion of ideas designed to help students develop a better understanding of the nature of proof in analysis. Topological concepts occur at various places in analysis and feel uncomfortable to many students. We look at several "levels" of topological concepts in Essay 10, and describe the purpose for introducing each of them into analysis.

Section 1. History and Biography

Essay 1. The Time of Archimedes and Other Giants (pre-1660)

It is helpful to think of the historical development of calculus as occurring in three time periods with different emphases. The first period begins with Archimedes of Syracuse and is an extended time of preparation for the discovery of calculus by Newton and Leibniz towards the end of the 1600s. The second period, characterized by an extensive development of results from the newly discovered calculus, extends from the discovery of the calculus by Newton and Leibniz in the years between 1665 to 1676 until the early 1800s. The final period, one of establishing a rigorous foundation of analysis to support the calculus, begins with Cauchy in 1821 and continues until almost the end of the nineteenth century. At the risk of oversimplification, we might say that calculus was discovered and developed during the second period and analysis was developed during the third period.[2]

Since a typical calculus course rarely includes material from the mathematics of the Greek period, one may be surprised to learn that many of our modern concepts can be traced back to the time during the third and fourth centuries B.C. The present-day definition of a real number comes from the theory of ratio and proportion by Eudoxus (408–355 B.C.), while results on the conic sections that are used throughout the calculus were developed by Apollonius (250–175 B.C.). Archimedes[3] (287–212 B.C.) was centuries ahead of his time with ingenious summing techniques to find areas of regions and volumes of solids of revolution (see Problem 27). In addition, the Greek insistence on rigor and proof, largely neglected until late in the nineteenth century, has come again into vogue to characterize the nature of modern mathematics, including analysis.

"If I have seen further than others, it is because I have stood on the shoulders of giants." This well-known quote attributed to Isaac Newton expresses the fact that many individuals developed results concerning areas, volumes, tangent lines, velocities, and infinite sums which Newton and Leibniz used as a guide for their discovery of calculus. In the rest of this essay, we present a brief account of the contributions made by some of these "giants."

By the time that Newton and Leibniz came on the scene, the stage was set for the discovery of calculus, by which we mean the fundamental relationship between the problem

[2]Robert L. Brabenec, *Introduction to Real Analysis*, PWS-Kent Publishing Company, Boston, 1990, pp. 21–25.

[3]For information about the work and time of Archimedes, see Chapter 4 in [**11**].

of finding slopes of tangent lines to curves and the problem of finding areas of regions bounded by curves. If these two men had not found this relationship, then likely one of their contemporaries would have. When the time is ripe for a new result, often more than one person makes the discovery independently of the others. This was the case with the discovery of non-Euclidean geometry in the early part of the nineteenth century by Gauss, Bolyai, and Lobachevsky.

We next list five of the contributions and conditions that were needed to prepare the way for the significant discoveries by Newton and Leibniz.

1. Improved methods of calculation
2. Appropriate symbolism and notation
3. Applications needing an explanation
4. Specific examples to motivate general results
5. A spirit of inquiry and curiosity among many individuals

After the death of Archimedes, there was virtually no progress of ideas associated with calculus until the invention of the printing press near the end of the fifteenth century led to a distribution of Greek mathematical writings in Europe.[4] In addition, the free spirit of inquiry during the Renaissance encouraged mathematical questions to be studied in earnest after an intellectual hibernation of some 1500 years. This era of scientific inquiry led to a revival of interest in matters such as planetary motion and navigation of the seas to seek new lands and ideas. Both of these required extensive mathematical calculations, and so better calculation techniques and devices were needed. It was long before the era of the hand-held calculator and the computer, but the discovery of logarithms by John Napier (1550–1617) provided a much needed tool for shortening calculations.

Another tool of great help was the development of better notation. Francoise Vieté (1540–1603), the French lawyer and court advisor to Henry IV, provided a significant improvement in algebraic symbolism and notation. One of his ideas was to distinguish between variables, which he represented by vowels, and parameters which he represented by consonants.

> The ability to deal with parameters as such focused attention on solution procedures rather than specific solutions themselves, and on questions concerning relationships between different problems. This shift in emphasis, from the particular to the general, was a necessary ingredient for the algorithmic approach that characterizes the calculus.[5]

In addition to this improvement in algebraic notation, the possibility of graphical representation of curves through the analytic geometry of René Descartes (1596–1650) blended together both algebra and geometry, providing a significant boost to intuition. The famil-

[4]While most history of calculus texts deal with European contributions, other cultures had their own historical development. For a treatment of the discovery of the series for arctan x in South India in the fifteenth century, see Ranjan Roy, "The Discovery of the Series Formula for π by Leibniz, Gregory and Nilakantha," *Mathematics Magazine*, Vol. 63, No. 5, December 1990, pp. 291–306.

[5]See p. 95 in [**12**].

iar saying that "one picture is worth a thousand words" is demonstrated in every calculus lecture when a function is graphically represented in the Cartesian plane.

While proper notation was an essential prerequisite for the discovery of the calculus, a wide variety of particular examples were needed as a basis from which to abstract general rules. As noted above, the first half of the seventeenth century was marked by questions about motion of various objects—the orbit of planets, the path of a projectile, or the descent of falling objects dropped from the Leaning Tower of Pisa. Johannes Kepler (1571–1630) considered the conjectures of Nicolas Copernicus (1473–1543) and the planetary calculations of Tycho Brahe (1546–1601) over a period of many years before summarizing his ideas in the three laws of planetary motion. He believed planets moved in elliptical orbits about the sun and that they traced out regions of equal area in equal periods of time. Newton would later prove these conjectures in his *Principia Mathematica.* In another very practical problem, Kepler developed several formulas to determine the amount of wine remaining in wine casks of various shapes. His methods previewed the idea of volumes for solids of revolution.

Many individuals developed various ways to calculate areas of regions, which offered improvement over the inscribed polygon method of Archimedes. These are all based on the method of dividing up the given region into small pieces and finding some way to determine a formula for the sum. A method often attributed to Cavalieri (see Problem 4) used rectangles with bases of equal length, while a method of Fermat (see Problem 16) used rectangles with base lengths forming a geometric sequence. The fact that the first general definition of the integral by Riemann would not be developed for another 200 years gives some indication of the difficulty of this problem.

Isaac Newton began his discovery of calculus via infinite series, and the binomial expansion in particular,[6] namely that

$$(a+b)^n = \sum_{i=1}^{n} \binom{n}{i} a^{n-i}\, b^i,$$

where the coefficients $\binom{n}{i}$ which represent the number of combinations of n things chosen i at a time are also the terms in the nth row of Pascal's triangle. John Wallis (1616–1703) from Oxford University used the principle of interpolation in his attempt to find the coefficients when the exponent n assumed such non-integer values as $\frac{1}{2}$ and $\frac{3}{2}$. Newton acknowledged his debt to Wallis for insights which helped some of his early work in finding the binomial series formula for arbitrary exponents.[7]

Although the derivative concept and its application to tangent lines to curves is taught in calculus courses before the integral concept and its application to areas and volumes, the historical order of development of these ideas was reversed. It was not until the 1630s that the tangent line question was addressed by individuals such as Fermat and Descartes.[8] Leibniz claimed it was while he studied the work of Pascal about the "characteristic tri-

[6] See Chapter 7 in [**11**].
[7] See pp. 271–7 in [**5**] for a presentation of Wallis's interpolation method.
[8] See pp. 122 ff. in [**12**].

angle"[9] that "a light suddenly burst upon him" and he saw the connection between finding tangent lines and finding the area of a region. Newton gained much insight while attending the lectures of Isaac Barrow at Cambridge University in the years 1664–5 about theorems connected with drawing tangents to curves and finding areas bounded by curves. These lectures were published as *Lectiones Geometricae* in 1670 and can be seen as the "culmination of the seventeenth century investigation leading towards the calculus."[10]

Essay 2. The Time of Newton and Leibniz (1660–1690)

If there is one fact students know about the historical development of calculus, it is that Isaac Newton (1642–1727) and Gottfried Wilhelm Leibniz (1646–1716) discovered the subject, and then fought over who discovered it first. Actually, the idea of finding slopes of tangent lines to curves and areas of regions had been around for several decades, if not centuries. Many individuals were successful in solving a variety of specific problems of these types by using several creative approaches (see Essay 1). What Newton and Leibniz discovered was that these two problems satisfy an inverse relationship, and their solutions can be expressed in terms of some general rules, which we today call differentiation and integration. This essay presents a comparison between the life and work of these two men.[11]

Three months after the death of his father, Newton[12] was born in the year when civil war broke out in England, which led to the Commonwealth under Oliver Cromwell—a time when England had no king or queen. The Aristotelian view of science was dominant in the universities, which served to discourage the growth of new ideas. A small group of intellectuals, who wanted the same freedom to experiment and find new results as Galileo had in Italy, began to meet in London. They were mostly from Oxford University and the mathematician John Wallis was one of the leaders. In 1660, they formed the Royal Society,[13] which is the oldest scientific society still in existence. In 1662, King Charles II granted a royal charter of incorporation and presented a royal mace, which is one of the treasures of the Royal Society until this day.

Newton's training was at Cambridge University, with strong interest in the experimental sciences. He devoted great effort to the study of optics and the nature of light. He also discovered results in physics, such as the law of gravitation and the laws of motion, being especially interested in applying them to planetary motion. He invented a reflecting telescope and was elected to the Royal Society in 1671 on the strength of this discovery.

The majority of his discoveries were concentrated within two 18-month periods of his life. The first was in 1665 and 1666 when Cambridge University was closed as a safety precaution with an outbreak of the black plague in London just 60 miles away. During

---

[9] See pp. 239–45 in [**12**].

[10] See p. 349 in [**6**].

[11] See [**4**] for a more complete version.

[12] There are many biographies of Newton with varying lengths. One of the more complete versions is Gale E. Christianson, *In the Presence of the Creator: Isaac Newton and His Times*, The Free Press, New York, 1984.

[13] E.N. da C. Andrade, *A Brief History of the Royal Society*, published by the Royal Society in 1960.

this time, Newton returned to his home in Woolsthorpe and made his discovery of the calculus, the law of gravitation, and the prism nature of light. The second creative period was in the mid 1680s when he again concentrated great effort to compose his *Philosophiae Naturalis Principia Mathematica* (Mathematical Principles of Natural Philosophy) which was published in 1687 under the encouragement of Edmund Halley (1656–1742).[14]

Leibniz also had an interest in the sciences, but his curiosity extended to all areas of knowledge. His early training was in law, philosophy, and languages, with schooling at the University of Leipzig. He developed new interests throughout his life, and sought for universal knowledge and competence. He earned his living by serving as a private secretary and researcher for two employers. First, he acted as legal advisor to the Prince Elector of Mainz for five years. After this employer's death in 1673, Leibniz began his service to the house of Brunswick, serving under three of its dukes until his death. The third duke became George I, King of England, in 1714.

Leibniz often traveled throughout Europe to carry out his duties. During the years 1672–76, he was in Paris on a mission to encourage King Louis XIV to wage war against Egypt in order to distract him from military action against the European countries. It was during this four year period that Leibniz began his study of mathematics and made his own discovery of calculus.[15] He made the acquaintance of Christiaan Huygens (1629–1695), Holland's greatest scientist who is especially remembered for his study of pendulum motion. In discussion with Leibniz about scientific matters, Huygens realized how deficient Leibniz was in his knowledge of mathematics and urged him to study the works of Blaise Pascal among others. As was his wont, when Leibniz began the study of a new field, he persisted until obtaining mastery. Huygens informed Leibniz of an unsolved problem in infinite series, namely to find the value for the sum of the reciprocals of the triangular numbers $1 + \frac{1}{3} + \frac{1}{6} + \frac{1}{10} + \cdots$. In work on this problem, Leibniz developed his harmonic triangle and the principle of telescoping sums (see Problem 2), which enabled him to show that the value of this series was 2. He later said that the principle behind this problem concerning the relation of the inverse operations of addition and subtraction led him to the idea of the relationship between the inverse processes of finding sums (integrals) and finding differences (derivatives). At this time, he also discovered the well-known result that[16]

$$\frac{\pi}{4} = 1 - \frac{1}{3} + \frac{1}{5} - \frac{1}{7} + \cdots.$$

During this four year period in Paris, Leibniz also traveled to London where he presented his model of a calculating machine to the Royal Society and became a member in 1673. He learned of Newton's work with what we now call calculus ideas.[17] Through the cooperation of the secretary of the Royal Society, Henry Oldenburg, Leibniz exchanged

---

[14]Halley was the individual who discovered Halley's comet.

[15]See Joseph E. Hofmann, *Leibniz in Paris 1672–1676: His Growth to Mathematical Maturity*, Cambridge University Press, Cambridge, 1974 for additional information.

[16]For an outline of Leibniz's proof, see pp. 264–6 in [**28**].

[17]Leibniz is actually the one who gave us this terminology with his terms *calculus differentialis* and *calculus integralis*.

ideas with Newton. Their relationship was cordial for twenty years until it was damaged by priority claims initiated largely by the English mathematicians John Wallis and John Keill.

In contrast to Leibniz's frequent travel, Newton lived a much more sedentary life, serving as a university professor in the Lucasian chair at Cambridge[18] for over twenty years. He assumed this position in 1669 when Isaac Barrow vacated it to become Chaplain to the King in London. In 1696, Newton left Cambridge to become first warden, then master of the mint in London until his death in 1727. During this time, he also served as President of the Royal Society from 1703 until his death.

Attention given to the question of priority of discovery at times seems to be greater than the attention given to the discovery of calculus itself. Here are the facts as best we know them today. Newton clearly made his discovery well before Leibniz did, probably in 1665. He wrote three treatises concerning his discoveries during the years 1669, 1671, and 1676, but these were not published until the years 1704, 1711, and 1736. The first treatise *De analysi per aequationes numero terminorum infinitas* was written in 1669 and contained his first version of calculus results. Although it was circulated among his acquaintances, it was not published until 1711. The second account *Methodus fluxionum et serierum infinitarum* was an enlargement of *De analysi* and contained the concepts of fluxions and fluents. After writing it in 1671, Newton was unable to find an interested publisher, so he lost interest and the work was not published until 1736, nine years after his death. The final monograph *De quadratura curvarum* was written in 1676 and published in 1704.

Newton also wrote two letters in 1676 to Leibniz through the intermediary Henry Oldenburg in response to Leibniz's request for information about Newton's discoveries. It did contain many of Newton's ideas, but the crucial idea was concealed by Newton in an anagram.[19] Leibniz was able to examine some of Newton's writings described above and copied parts to take with him after he left London and Paris to return to Leipzig.

Although Leibniz was almost ten years behind Newton in his discovery of calculus, he was the first to publish details of these results.[20] This occurred in 1684 in an issue of the journal *Acta Eruditorum* which Leibniz founded in Leipzig in 1682. The title of this 6-page article is "A New Method for Maxima and Minima, as Well as Tangents, Which Is Not Obstructed by Fractional or Irrational Quantities." The writings of Leibniz were not easily understood, and it was left to James and John Bernoulli, the Marquis de L'Hospital, and Leonhard Euler to clarify and extend his ideas (see Essay 3).

When the priority question was finally formally examined, the cards were clearly stacked against Leibniz, since the commission was formed by the Royal Society of London and Newton was the presiding officer. It should be no surprise then that the case was decided in Newton's favor. His reputation continued to rise, and after his death, he was buried in Westminster Abbey. By contrast, Leibniz died virtually alone and forgotten in 1716, never receiving well-deserved accolades and recognition for his many achievements. In addition to those already described, Leibniz founded the Berlin Academy in 1700. The prestige of this scientific academy was enhanced by the presence of Leonhard Euler from

---

[18] Stephen Hawking now holds this position. He is the author of *A Brief History of Time*.

[19] The actual anagram was $6accdae13eff7i3l9n4o4qrr4s8t12vx$.

[20] Unless you consider the details Newton put into his two letters in 1676.

1741 to 1766 and that of Joseph Lagrange from 1766 to 1787. The University of Berlin was founded in 1809 and occupied a position of prominence in mathematics during the last half of the nineteenth century. Leibniz's efforts to work on a universal language for calculation and decision-making provided an important link in the chain of contributions leading to the role of computers in our present society. So with the passage of time, we can better see the value of his work.

Essay 3. The Time of Euler and the Bernoullis (1690–1790)

After Newton and Leibniz discovered the fundamental relationship between the derivative and integral concepts, it was largely the efforts of the following three men born in Basel, Switzerland to develop calculus into the useful collection of methods, formulas, and applications that we have available today.

James Bernoulli	1654–1705
John Bernoulli	1667–1748
Leonhard Euler	1707–1783

James and John Bernoulli were brothers from a large family. They worked hard to understand the ideas of Leibniz about calculus and were able to express these ideas in a simpler form which led to a popularization of calculus in continental Europe during the 18th century. By contrast, the countrymen of Newton—such as Brook Taylor and Colin Maclaurin—could not accomplish this with the equally complicated fluxion terminology of Newton. John Bernoulli had two sons who were also gifted mathematicians, as well as contemporaries and friends of Leonhard Euler. There are many other connections between these three men.

James Bernoulli was about thirty years old in the mid 1680s when he was introduced to the ideas of calculus through some of the papers written by Leibniz. During the next 20 years before his relatively early death, he wrote several articles on calculus which were published in the journal *Acta Eruditorum* that was begun by Leibniz. James investigated the properties of many of the famous curves of interest in his day, including the catenary, tractrix, cycloid, logarithmic spiral, and the lemniscate. He also developed many results about probability and collected them in a book *Ars Conjectandi* (The Art of Conjecturing) which was published eight years after his death. The concept of Bernoulli numbers, which occur often in analysis, are found in this book (see Problem 30).

John Bernoulli was thirteen years younger than his brother James, and he lived 43 years past his brother's death. While both brothers spent most of their careers in the study of calculus, there were several differences in the nature of their achievements. We already noted that John lived thirty years longer than James. He married and had three sons with mathematical ability. His oldest son Nicholas died by drowning at the age of thirty, while his middle son Daniel lived a long life as a productive mathematician and teacher. John was himself a professor of mathematics at the University of Basel for most of his life, assuming the position left vacant when his brother James died in 1705. His most famous student was Leonhard Euler. An earlier student was a French nobleman whom he tutored privately. This was the Marquis de L'Hospital who had mathematical ability and a desire to learn about the new discoveries of the calculus (see Selection 1 in Part IV).

Leonhard Euler[21] is considered one of the greatest mathematicians who ever lived. He certainly was the most prolific writer with over 500 books and papers estimated to fill perhaps 80 large volumes when the publication of his works is finally completed. By comparison, the men who are usually mentioned as the three greatest mathematicians—Archimedes, Newton, and Gauss—produced relatively few mathematical publications. Euler lost the sight in one eye at an early age and became totally blind by the age of 60, yet his mathematical productivity never slowed. He was not only supremely competent in the field of calculus with its associated computations, but also in the more abstract field of number theory, and in applied mathematics as well. He wrote several textbooks in a very lucid form, introducing many of the standard results and notation that are still used today. He fathered 13 children and took time to care for their moral instruction. Some of his children wrote out his discoveries after Euler became blind. His life is best viewed by considering its division into the following four time periods.

Period 1: In Basel, Switzerland 1707–1726 (ages 0–19)

Period 2: At St. Petersburg Academy 1726–1741 (ages 19–34)

Period 3: At Berlin Academy 1741–1766 (ages 34–59)

Period 4: At St. Petersburg Academy 1766–1783 (ages 59–76)

During his youth in Basel, Euler pursued a broad field of study. His father was a pastor and wanted his son to follow him in that profession. Euler was given the opportunity on Saturday mornings for private tutoring from John Bernoulli, who was then the mathematics chair at the University of Basel. He recognized the mathematical talent of Leonhard, and successfully encouraged his father to permit Leonhard to pursue a career in mathematics.

Peter the Great founded a scientific academy in St. Petersburg which he hoped would help Russia compete with the rest of Europe in scientific matters. After Peter's death in 1725, his wife Catherine tried to carry out his plan. Two sons of John Bernoulli were offered positions at this academy, and quickly secured an invitation for Leonhard to follow them. Though not yet twenty years of age, Euler left his home country forever to accept a position in the physiology section of the St. Petersburg Academy, as there were no openings in mathematics at that time.

Shortly after this decision was made, Catherine died and Daniel Bernoulli moved back to Basel. Leonhard changed his affiliation to the mathematics department, married, and began work on his career. This is the time period when he discovered the famous result that

$$\sum_{n=1}^{\infty} \frac{1}{n^2} = \frac{\pi^2}{6}.$$

[21]There is no full biography of Euler. For biographical information, see pages 139–152 in [**2**], pages xix–xxviii in [**10**], or pages 736–742 in [**16**].

This discovery delighted John Bernoulli and amazed James Stirling.[22] Hardly less amazing was the related result that

$$\sum_{n=1}^{\infty} \frac{1}{n^{26}} = \frac{2^{24} \cdot 76377927 \cdot \pi^{26}}{27!},$$

and the fact that a Bernoulli number is part of this formula. During this 15 year period at St. Petersburg, Euler's reputation grew enough that Frederick the Great invited him to come to the Berlin Academy. Since the political situation was quite unstable in Russia in 1741, Euler was glad for an excuse to leave, and he accepted the Berlin offer.

Actually, Euler was not the first choice of Frederick the Great, who was a great admirer of the French Enlightenment and was enamoured with Jean le Rond d'Alembert (1717–1783). This mathematician was coauthor with Denis Diderot of the 28-volume *Encyclopédie*, written between 1751 and 1772 and containing the new insights in science and the arts during the Enlightenment.[23] However, d'Alembert refused the overtures of Frederick the Great, so Euler became the second choice. Euler was clearly not a figure in the spirit of the Enlightenment. One of his non-mathematical works, *Letters to a German Princess*, was a collection of brief essays written to give instruction to a real princess about scientific matters, often with comments based on Euler's religious beliefs. It proved to be a very successful book, and was translated into many languages. However d'Alembert and Lagrange felt Euler compromised his scientific standing by writing this book.[24]

When Euler moved his young family to Berlin to spend the next quarter of a century, he was already blind in one eye. This was the period when he wrote his great calculus textbooks, such as the *Introductio in analysin infinitorum* in 1748. Throughout his entire time in Berlin, Euler continued to receive a salary from the St. Petersburg Academy and he in turn continued to send them mathematical papers to publish in their journal. He also wrote letters to many correspondents, including Lagrange who was then teaching at a military school in Turin, Italy. Eventually Euler tired of Berlin and tensions with Frederick the Great. When Catherine the Great made repeated offers for him to return to St. Petersburg, Euler arranged for Lagrange to replace him at Berlin and moved back to St. Petersburg. During the time of this move, Euler became completely blind and yet he continued to be a productive mathematician during the final 17 years of his life.

We conclude this essay with a summary of contributions made by some "lesser lights" during this time period. The Marquis de L'Hospital (1661–1704) wrote the first calculus text in 1696. Although limited to differential calculus, it was a worthy contribution to the mathematical literature of its time. We remember that although Newton and Leibniz and James and John Bernoulli learned about calculus before the Marquis, he was the first to organize the material into book form.

---

[22] See the letters between Euler and Stirling in Selection 2 in Part IV.

[23] See Thomas L. Hankins, *Jean d'Alembert: Science and the Enlightenment*, Clarendon Press, Oxford, 1970.

[24] George Sarton, "Lagrange's Personality," *Proceedings of the American Philosophical Society*, Vol. 88, No. 6, 1944, p. 477.

While the Bernoullis and Euler championed the Leibniz form of the calculus, Brook Taylor (1685–1731), James Stirling (1692–1770), and Colin Maclaurin (1698–1746) were the countrymen of Newton who used the fluxion and fluent terminology in their treatment of calculus. Their names are familiar to us today in connection with the Maclaurin series and Taylor series representation of functions. Taylor's formula occurred in his 1715 book titled *Methodus incrementorum directa et inversa.* Maclaurin's *Treatise of Fluxions* was published in 1742, and was intended as a response to the attack of Bishop Berkeley in 1734 upon the weaknesses of fluxions as a foundation for the calculus. The following subtitle of Berkeley's book describes the nature of his challenge.

> On a Discourse Addressed to an Infidel Mathematician Wherein It Is Examined Whether the Object, Principles, and Inferences of the Modern Analysis are More Distinctly Conceived, or More Evidently Deduced, than Religious Mysteries and Points of Faith. "First Cast the Beam Out of Thine Own Eye: and Then Shalt Thou See Clearly to Cast Out the Mote Out of Thy Brother's Eye.

He was essentially correct, and left mathematicians with a foundational problem that took almost a century to correct. As mentioned, Maclaurin's book was a first response to Berkeley, but it was inadequate. D'Alembert called for a theory of limits as the answer, but was unable to supply details for this. Lagrange felt the use of power series was the correct approach for a foundation. In the next selection, we see the limit concept was the correct path, and that Cauchy was able to develop it in a very adequate manner.

Essay 4. The Time of Cauchy, Fourier, and Lagrange (1790–1850)

Mathematically speaking, France was the place to be during the first half of the 17th century. This was the time and place of Vieté, Descartes, Pascal, and Fermat, and their collective contribution to mathematics exceeded that from any other country during this time. However, after this period, except for a few isolated examples such as the Marquis de L'Hospital and Jean Le Rond d'Alembert, French mathematicians made little impact for more than a century. All this changed dramatically near the end of the 18th century. After the French Revolution in 1789, Napoleon Bonaparte assumed control of France as emperor and military leader. He believed his officers needed to be trained in mathematics and science, so he founded the École Polytechnique to provide this training and also founded the École Normal to train teachers of science and mathematics. As a consequence, Paris again became the center of the mathematical world during the first half of the 19th century, and three of the most influential analysts of this period were Cauchy, Fourier, and Lagrange.

Augustin-Louis Cauchy[25] (1789–1857) was born in the year of the revolution and educated at the newly formed École Polytechnique. After graduation, he served as a mining engineer and then returned to teach mathematics at the École Polytechnique. While others before him had been concerned about the lack of rigor in calculus and had tried various ways to correct this, Cauchy was the first to show how the limit concept could serve as the foundation for calculus. He did this through his lectures and his book *Cours d' analyse*

[25]See Bruno Belhoste, *Augustin-Louis Cauchy: A Biography*, Springer-Verlag, New York, 1991 for biographical information.

that was published in 1821. Cauchy was a prolific writer, second only to Euler, with 789 papers filling 26 volumes of his collected works. At times he submitted enough articles to fill whole issues of a journal, and the Academie des Sciences had to impose a four page limit on articles for the *Comptes Rendus* in order to give other writers a chance to publish.

The content and structure of our present-day real analysis course is due mainly to Cauchy's influence. Although the familiar ϵ-δ notation itself did not originate with Cauchy, the idea to base the other concepts of calculus—continuity, the derivative, the definite integral, sequences and series—on the concept of a limit was his important contribution. His definition of limit below shows how the idea was then expressed in words, rather than the symbolic form that is familiar today.

> When the values successively attributed to a particular variable approach indefinitely a fixed value, so as to finish by differing from it by as little as one wishes, this latter is called the limit of all the others.

There were weaknesses in his original work of the 1820s that needed to be corrected by others, but this should not detract from the significance of Cauchy's contribution. One example of a weakness is his failure to see the need for the concept we refer to today as uniform convergence. One of Cauchy's theorems stated that the limit of a convergent sequence of continuous functions must be a continuous function. Abel knew this was incorrect and presented a counterexample, but Cauchy was reluctant to accept it. It was Paul Seidel who recognized around 1848 the need for a concept of uniform convergence to prove such a theorem.

Cauchy exhibited some unusual personal traits. He was firmly devoted to the king of France and went into voluntary exile when the king was deposed after the 1830 revolution. He refused to take any oaths of allegiance that were required in order to teach at the university (his financial situation was such that he could live comfortably without a salary from the state). He had an annoying habit of misplacing the papers of younger mathematicians, such as Evariste Galois (1811–1832) and Niels Abel (1802–1829), who had sent their work to the Paris Academy for evaluation. Part of a letter from Abel to his friend Holmboe in 1826 includes the following assessment:

> Cauchy is a fool and one can't find any understanding with him, although he is the mathematician who at this time knows how mathematics should be treated ... he is extremely catholic and bigoted ... I have worked out a large paper on a certain class of transcendental functions to present to the Institute ... I showed it to Cauchy, but he would hardly glance at it.

Our second great mathematician, Joseph B. Fourier[26] (1768–1830), was born into a working class home and orphaned at an early age. Without the new equality brought about by the French Revolution, he would not have had the chance to enter scientific studies, and mathematics would be the poorer for it. Even so, he had limited opportunity for advanced

[26] See John Herival, *Joseph Fourier: The Man and the Physicist*, Clarendon Press, Oxford, 1975 for biographical information.

study before being chosen by Napoleon to accompany him on the Egyptian campaign. He was given administrative opportunities and also wrote the introduction for the *Description of Egypt* which has been cited for its literary excellence. His *Theory of Heat* was called a "mathematical poem" by Lord Kelvin and was chosen as one of only three mathematical works to be included in the Great Books of the Western World series from the University of Chicago.

Fourier made quite a different contribution to mathematics than did Cauchy. He was more a physicist than a mathematician. His main book *Theorie analytique de la chaleur* presented a mathematical solution to the partial differential equation for heat flow. It was suggested by another problem—that of the vibrating string—which was studied in the time of Euler and the Bernoullis. In his book, Fourier developed the trigonometric series that now bear his name. Power series had been used effectively since the time of Newton to represent the trigonometric and exponential functions and their inverses, but they could not represent the discontinuous functions that were needed for these new problems.

By the end of the 18th century, it was thought there were no more problems left to be solved by the tools of calculus. Lagrange expressed this belief when he wrote in a 1781 letter to d'Alembert:

> It seems to me that the mine is already almost too deep, and unless we discover new seams we shall sooner or later have to abandon it. Today physics and chemistry offer more brilliant and more easily exploited riches; and it seems that the taste of the century has turned entirely in that direction. It is not impossible that the mathematical positions in the academies will one day become what the university chairs in Arabic are now.

Fourier series provided a way to represent discontinuous functions and thereby opened up new vistas for calculus throughout the 19th century. In fact, while Georg Cantor was working on the convergence problem for Fourier series around 1870, he began his famous and very influential work on sets, which led to a revolution in thinking about the infinite. But Fourier did not find the mathematical world of his day waiting for his results with open arms. He originally submitted his work to the Paris Academy in 1807. The committee assigned to evaluate it, consisting of Lagrange, Laplace, and Legendre, apparently did not understand these new ideas because they rejected the paper. It was not until 1822 after extensive revisions that this work was printed, and Fourier finally began to receive the recognition he had long sought.

Our final analyst, Joseph-Louis Lagrange (1736–1813), was the youngest of eleven children and the only one to survive past infancy. Even though he was Italian by birth, he is closely associated with the contingent of French mathematicians that was active at the end of the 1700s and the beginning of the 1800s. The reason for this is that after spending his early years at the University of Turin and an intermediate period from 1766 to 1786 at the Berlin Academy, Lagrange spent the last 27 years of his life in Paris. In this setting, he was widely respected as an elder statesman of mathematics.

Along with other contemporaries, such as Legendre and Cauchy, Lagrange organized his lecture notes into influential textbooks that played a significant role in the dissemination of calculus results. He consistently called for a greater level of rigor in calculus and believed it could be achieved by using the power series concept as a basic foundation. He developed this idea in his *Théorie des fonctions analytiques* which was published in 1797. Within 24 years, Cauchy would follow this example of basing calculus concepts and results on a careful foundation, but he chose the limit concept instead of power series for this purpose.

There were many other French mathematicians during this time, such as Gaspard Monge (1746–1818), Pierre-Simon Laplace (1749–1827), Adrien Marie Legendre (1752–1833), L. M. N. Carnot (1753–1823), J. V. Poncelet (1788–1867), and Joseph Liouville (1809–1882), but their contributions lay elsewhere than with the development of real analysis. For example, Pierre-Simon Laplace, sometimes called the Newton of France, wrote two multi-volumed works. One dealt with celestial mechanics, showing how to apply calculus to obtain Newton's laws, and the other was concerned with probability.

This period was also the time when Carl Friedrich Gauss[27] (1777–1855)—the man usually considered the greatest mathematician of all time—lived in what is today called Germany. Because of his importance to mathematics, we include some information about him even though his main contributions were in such fields as number theory, geometry, and astronomy, rather than in calculus and real analysis. At the age of 18, he discovered a proof that the regular polygon of seventeen sides could be constructed using only compass and straightedge. He kept a diary for many years in which he recorded cryptic notes about his main discoveries. In 1801, his most famous book, the *Disquisitiones arithmeticae*, was published. This book began with a presentation of the congruence relation and launched the modern theory of numbers. Gauss published this small book himself after the French Academy rejected it in 1800. Unlike Cauchy and Lagrange, Gauss had a dislike for teaching students and writing textbooks. He kept secret many of his discoveries because he wanted to avoid controversy. There is quite a long list of disgruntled mathematicians who announced their work only to have Gauss produce evidence that he had developed the idea years before. Some examples of this include the discovery of non-Euclidean geometry by Janos Bolyai (1802–1860) and Nicolai Lobachevsky (1793–1856), results on elliptic functions and integrals by Niels Abel (1802–1829) and Carl Gustav Jacobi (1804–1851), and the geometric representation of complex numbers by J. R. Argand (1768–1822) and C. Wessel (1745–1818). Gauss did make one important contribution to the development of real analysis through a memoir in 1813 in which he was the first to develop a rigorous treatment of convergence when he discussed convergence of the hypergeometric series. But he did not carry these efforts further.

Niels Abel is known today as the Norwegian mathematician who discovered the impossibility of solving the quintic equation. But he and Cauchy also receive credit for developing many of the standard convergence tests for infinite series. In addition, David Bressoud described his surprise to see Abel and Peter Gustav Lejeune Dirichlet (1805–1859) as "central figures of the transformation of analysis that fits into the years from 1807 through 1872."[28]

---

[27] See W. K. Búhler, *Gauss: A Biographical Study*, Springer-Verlag, New York, 1981 for biographical information.

[28] See p. ix in [**5**].

Bernard Bolzano (1781–1848) from Prague was a remarkable individual who was far ahead of his time in many ways. He developed such results from analysis as a proof for the intermediate value theorem and some properties of limit points for a sequence. His best known book is *Paradoxes of the Infinite*. However, he was viewed more as a churchman and philosopher than a mathematician, and most of his results were not published until after his death, so he did not have much influence on the development of calculus. His name has been preserved, however, in the Bolzano–Weierstrass theorems.

Essay 5. The Time of Riemann and Weierstrass (1850–1900)

During the 1790–1850 time period discussed in Essay 4, the mathematical world was centered in Paris at the Académie des Sciences and the École Polytechnique, founded by Napoleon as part of his plan for military dominance in Europe. He believed his officers would be best prepared by a strong education in mathematics and the physical sciences. This lesson was not lost on Wilhelm von Humboldt (1767–1835) of Prussia, who accurately saw his country as a non-unified collection of states, vunerable to the attacks of Napoleon. Von Humboldt initiated a plan in the 1820s to unify and strengthen Germany by reforming the educational system of Prussia and establishing strong universities.[29] So successful was this plan that in the Franco-Prussian War of 1870, Germany was able to quickly defeat France (and the second Napoleon). The following list of German mathematicians of prominence during the nineteenth century, in comparison with hardly any during previous centuries serves as additional evidence of the success of this plan. This list includes Carl Friedrich Gauss (1777–1855), Carl Gustav Jacobi (1804–1851), Peter Lejeune Dirichlet (1805–1854), Ernst Kummer (1810–1893), Karl Weierstrass (1815–1897), Leopold Kronecker (1823–1891), Georg Bernhard Riemann (1826–1866), Richard Dedekind (1831–1916), Georg Cantor (1845–1918), Gottlob Frege (1848–1925), Felix Klein (1849–1925), and David Hilbert (1862–1943).

There were relatively few faculty positions in German universities during the nineteenth century, so competition was intense for some of the more coveted positions. Because of the great prestige that Gauss brought to the University of Göttingen during the years from his initial appointment there in 1807 until his death in 1855, it was a great honor to be chosen as one of his successors. Dirichlet was the first successor, but he only held the position for four years until his death in 1859. Riemann was the next successor, but his tenure was also cut short by his untimely death to tuberculosis in 1866. David Hilbert came to Göttingen in 1895, and after 1900, he made Göttingen the most prestigious center in the world for mathematical study during the early part of the twentieth century, until its influence was destroyed by the Nazi regime in the 1930s.

A second university of prominence was the University of Berlin. The Berlin Academy had been founded by Leibniz in 1700, and had two great mathematicians located there in the eighteenth century—Leonhard Euler from 1741 to 1766 and Joseph Lagrange from 1766 to 1787. The University of Berlin was founded in 1807. Dirichlet and Jacobi were two prominent faculty there during its early years. At its zenith of influence during the

[29] Carl B. Boyer and Uta C. Merzbach, *A History of Mathematics*, Second edition, John Wiley & Sons, 1989, p. 593.

last half of the nineteenth century, the university had three great mathematicians at one time—Kummer and Kronecker in algebra and Weierstrass in analysis.

The experience of Georg Bernard Riemann[30] illustrates the typical career path for a mathematician in nineteenth-century Germany. He began his studies at the University of Berlin and then moved to the University of Göttingen where he received his doctorate in 1851. Gauss was one of his teachers, but Riemann was not impressed by his abilities as a teacher. His doctoral thesis was on the subject of a general theory for functions of one complex variable. Cauchy initiated the serious study of such functions, and Riemann and Weierstrass completed the general theory as part of their contribution to analysis.

At this time, Riemann was given the "privilege" of teaching as a Privatdocent, but this position brought no salary. There were two final hurdles to clear before Riemann could be granted the position of assistant professor of mathematics and receive a salary for his work. The first, the "habilitationschrift," consisted of a series of lectures before the university council. Riemann presented these in December 1853 on the topic of representing functions by trigonometric series. The final requirement, the "habilitationsvortag," was a report on a more narrow topic, chosen by the university council from a list of three possibilities submitted by the candidate. In light of his previous research in complex numbers and trigonometric series, the first two choices were easy for Riemann to make. He was sure that one of these would be chosen, but since he needed a third option, he chose the field of geometry. For some reason, Gauss, who was now in the last year of his life, persuaded the council to choose this final topic. Riemann unfortunately had only a title but no results, so he needed to pursue a crash course in geometry. His presentation "On the Foundations of Geometry" in June 1854 was praised by Gauss and is considered one of the most outstanding short mathematical papers ever written.[31]

A student of analysis encounters the name of Riemann in the unit on integration. For almost 200 years before Riemann, integration was considered the same as antidifferentiation. That is, a function $f(x)$ was considered to be integrable if one could find an antiderivative for $f(x)$. Riemann's definition for the integral of a function made it possible to decide whether or not a function was Riemann integrable, independent of whether an antiderivative could be found. He also presented the Riemann integrability criterion, a necessary and sufficient condition for a function to be integrable, and the result needed to prove many integration theorems. The power of this criterion was enhanced by the fact that one need only consider partitions of the original interval into subintervals of equal length.

One final comment about Riemann. Now that Fermat's last theorem has been proven, most mathematicians consider a conjecture made by Riemann to be the most important unsolved problem in mathematics. This conjecture, called the Riemann hypothesis, deals with the location of the zeroes of the zeta function, which is defined over the domain of complex numbers. When a real number $p > 1$ is chosen as a domain value, the zeta

---

[30] See the biographical information in Michael Monastyrsky, *Riemann, Topology, and Physics*, 2nd edition, translated by James and Victoria King and edited by R. O. Wells, Jr., Birkhäuser, Boston, 1999.

[31] An English translation of this paper is in Willliam B. Ewald, ed., *From Kant to Hilbert: A Source Book in the Foundations of Mathematics*, Clarendon Press, Oxford, 1996, pp. 652–651.

function assumes the range value of

$$\sum_{n=1}^{\infty} \frac{1}{n^p}.$$

The career path of Karl Weierstrass[32] is all the more unusual when considered against the backdrop of Riemann's experience. Weierstrass was a less than dedicated student, who failed to complete a regular course of university study. He settled for certification in education, and spent almost twenty years teaching at the high school level. During this time, he also pursued significant mathematical research, often working through the night hours. When one of his papers on the theory of Abelian functions was published in *Crelle's Journal*[33] in 1854, it made such an impression on the mathematical community that he was given an honorary doctorate by the University of Königsberg on March 31, 1854. When a second paper, containing even more significant results on Abelian functions, was published in *Crelle's Journal* in 1856, Weierstrass was appointed to the staff at the Industry Institute in Berlin, made a member of the Berlin Academy, and offered an assistant professorship at the University of Berlin. Over the next three decades, Weierstrass became one of the most respected teachers of mathematics. He was also responsible for the following significant contributions to analysis and the development of its rigor.

1. He presented a rigorous development of the real numbers through his "arithmetization of analysis" program. Dedekind also contributed to this effort with the use of "cuts" of rational numbers to represent irrational numbers.
2. He presented outstanding lectures on analysis which were copied and distributed by his students. He himself published relatively few papers.
3. He was one of three mathematicians, Cauchy and Riemann being the other two, who developed the field of complex analysis.
4. His examples, such as the one for an everywhere continuous yet nowhere differentiable function, demonstrated how faulty intuition could be in comparison to careful rigor.

In 1870, Weierstrass met Sonja Kovalevsky[34] (1850–1891), who had come to Germany from Russia because women were not allowed to study in the universities of Russia. The situation was not much better in Germany, but Weierstrass, at the age of 55, agreed to privately tutor this twenty-year-old woman. Sonja claimed to have learned calculus from the large sheets of a calculus text which her father used as wallpaper in her room. She wrote a memoir titled *Memories of a Russian Childhood.* Weierstrass regarded her fondly, somewhat as a daughter, but she often appeared indifferent to his gestures of friendship.

[32]See Chapter 22 in [**2**] for more detail of the life and work of Weierstrass.

[33]*Crelle's Journal* was started in 1826 by the German engineer August Crelle to encourage the spread of mathematical knowledge in Germany. Crelle befriended the young Norwegian mathematician Niels Abel and published several of his papers during the early years of this journal.

[34]The following two biographies of Sonja Kovalevsky also contain interesting information about Karl Weierstrass. Don H. Kennedy, *Little Sparrow: A Portrait of Sophia Kovalevsky*, Ohio University Press, Athens, Ohio, 1983 and Ann Hibner Koblitz, *A Convergence of Lives Sofia Kovalevskaia: Scientist, Writer, Revolutionary*, Rutgers University Press, New Brunswick, New Jersey, 1993.

He was able to secure a teaching position for her in Sweden, where she died in 1891 at the age of 41. Weierstrass died in 1897 at the age of 82.

It is interesting to note that male mathematicians often were supportive of the few females in mathematics prior to the last half of the twentieth century. Gauss praised the work of Sophie Germain when she first wrote under a male pseudonym, and was even more encouraging when he learned who she really was. David Hilbert supported the opportunity for Emmy Noether to teach at the University of Göttingen in the 1920s against the opposition of the school administration.

We conclude this essay by mentioning the contributions of the man whose discoveries paved the way for a significant expansion and generalization of mathematics, and analysis in particular, during the start of the twentieth century. Georg Cantor specialized in the study of analysis as a student of Weierstrass at the University of Berlin. His research interest was concerned with sets of convergence for Fourier series of various functions. Since these sets were so different from, and more complicated than, the intervals of convergence for power series representation of functions, Cantor took time aside from analysis to make a general study of sets, and especially infinite sets. The rest of the story is history, as Cantor discovered his revolutionary results about the nature of infinite sets, and spent the rest of his career on this topic.[35] To him, we owe the distinction between countably infinite sets and uncountable sets, as well as the unending hierarchy of cardinal and ordinal numbers. The continuum problem of Cantor became the first one on Hilbert's famous list of 23 problems which he presented to the International Congress of Mathematicians in Paris in 1900. These problems determined the course of much research in the twentieth century.[36] The set theory developed by Cantor became the basis for a generalized theory of measure and integration.

Cantor also played a significant role in the formation of a German association of mathematicians to sponsor regular conferences where mathematics could be discussed. This was the beginning of a new method to communicate and disseminate mathematical knowledge. Within a few years, the American Mathematical Society and the Mathematical Association of America were formed, and the International Congress of Mathematicians first met in 1897.[37] These organizations and their meetings continue to influence the development of mathematics to the present day.

Essay 6. The Time of Lebesgue and Hardy (1900–1940)

Two mathematicians, Henri Poincaré (1854–1912) at the University of Paris and David Hilbert[38] (1862–1943) at the University of Göttingen in Germany were likely the most dominant mathematicians during the opening years of the twentieth century. David Hilbert

[35] See Joseph Warren Dauben, *Georg Cantor: His Mathematics and Philosophy of the Infinite*, Harvard University Press, Cambridge, Massachusetts, 1979.

[36] See Benjamin H. Yandell, *The Honors Class: Hilbert's Problems and Their Solvers*, A K Peters, Natick, MA, 2002 for some interesting reading.

[37] For an interesting description of these congresses from 1897 until 1986, see the book Donald J. Albers, G. L. Alexanderson, Constance Reid, *International Mathematical Congresses: An Illustrated History*, Springer-Verlag, New York, 1987.

[38] See Constance Reid, *Hilbert*, Springer-Verlag, New York, 1970 for an interesting biography.

set the tone for the century through his lecture at the International Congress of Mathematicians in Paris in August, 1900, in which he presented 23 problems as a challenge for mathematicians in the new century. It is a measure of the influence of Hilbert that these problems motivated much mathematical research during the entire twentieth century. During the years of the growth of various philosophies of mathematics in the early part of the twentieth century, Hilbert developed the formalist school to counter the influence of the intuitionist school of Kronecker, L. E. J. Brouwer (1881–1966), and Poincaré and of the logicist school of Gottlob Frege and Bertrand Russell. Hilbert worked for about 30 years to develop his program before the incompleteness theorem of Kurt Gödel in 1930 showed that the main goal was unattainable.

Henri Poincaré is often considered the last universal mathematician, in that he was knowledgeable of the developments in practically every branch of mathematics in his day. His special strengths were in astronomy and mathematical physics, the theory of functions, algebraic topology, and differential equations. In his later years, he developed a reputation as a popularizer of mathematics.

> With the growth of his international reputation, Poincaré was more and more called upon to speak or write on various topics of mathematics and science for a wider audience, a chore for which he does not seem to have shown great reluctance. His vivid style and clarity of mind enhanced his reputation in his time as the best expositor of mathematics for the layman. His well-known description of the process of mathematical discovery[39] remains unsurpassed and has been on the whole corroborated by many mathematicians, ...[40]

Though neither Hilbert nor Poincaré played a significant role in the development of analysis, we included some comments about them because of their dominant position in the mathematical world in the early 1900s, similar to the role of Gauss in the early 1800s.

In the previous two essays, we saw how the level of rigor in analysis increased significantly during the 19th century. It began with Cauchy's introduction of the limit concept in 1821 as a workable basis for defining the important concepts of calculus. The results of uniform continuity and uniform convergence were added along with a careful definition of the Riemann integral by the year 1860. Then Weierstrass and others provided a careful, axiomatic development of the real numbers and their properties, based on those of the natural numbers. Weierstrass's strange example of an everywhere continuous, yet nowhere differentiable function, underscored significant differences between continuity and differentiability.

Then Cantor's introduction and development of the concept and properties of infinite sets began a whole new era of generalization and abstraction in analysis. The variety of subsets of reals that could now be considered in analysis increased dramatically, along with attempts to generalize the length of interval concept to define a measure for these new

[39]One source for this article titled Mathematical Creation is in James R. Newman, editor, *The World of Mathematics*, Simon and Schuster, New York, 1956, pp. 2041–50.

[40]See p. 2022 in [**16**].

sets (see Problem 33). The definition of measure presented by Henri Lebesgue[41] (1875–1941) in 1902 became the prototype, leading to the Lebesgue integral which generalized the Riemann integral and corrected some of its weaknesses.

Another development issuing from Cantor's theory of sets was the subject of point set topology. It evolved during the first twenty years of the twentieth century, furnishing the language and techniques to prove such important results as the theorems about continuous functions (see Essay 10). The introduction of general spaces, such as topological spaces and linear vector spaces provided the framework for such future generalizations as that of functional analysis. Felix Hausdorff (1868–1942) played a role in the development of set-theoretic topology when he provided an axiomatic basis for a topological space, stated in terms of neighborhoods in his book *Grundzüge der Mengenlehre* (Basic Features of Set Theory), which was published in 1914.[42] Stefan Banach (1892–1945) represented the influential Polish school in presenting properties of the normed linear vector spaces which bear his name today.

So Lebesgue played an important role in the opening years of the twentieth century, not just for his discovery of Lebesgue measure and integration, but also for his encouragement of a significant move towards abstraction. This movement would perhaps reach its ultimate after 1940 with the efforts of a group of French mathematicians, who worked collectively under the pen name of Nicolas Bourbaki, to carefully rewrite all mathematics, including analysis, using the new language and rigor of the twentieth century.[43]

The other main figure in this time period is G. H. Hardy (1877–1947). He is of interest to us for several reasons. For one, his work marked the reemergence of analysis in England after a self-imposed hibernation of more than 150 years. Newton may have won the initial battle in his skirmishes with Leibniz about who first discovered the calculus. But Leibniz, through the efforts of his followers such as James and John Bernoulli and Leonhard Euler, won the war since the differential notation of Leibniz was significantly more popular and useful than the fluxion notation of Newton and his countrymen. In response to this, after the death of Maclaurin in 1748, British mathematicians neglected analysis and turned to the field of algebra where they were ready to make strong contributions in the nineteenth century through the efforts of such men as George Peacock (1791–1858), William Hamilton (1805–1865), Augustus DeMorgan (1806–1871), James Joseph Sylvester (1814–1897), George Boole (1815–1864), and Arthur Cayley (1821–1895). This British abstinence from analysis changed as Hardy devoted himself to a long career at the Universities of Cambridge and Oxford, writing many papers and books on the subject of analysis. Some of his book titles include *A Course of Pure Mathematics*, *Divergent Series*, and *A Mathematician's Apology*. This last book is a brief, but poignant, statement about the forces that drive mathematicians in their research. The following quote is from this book.

---

[41] See Selection 4 in Part IV for a brief biography of Lebesgue.

[42] Ioan James, *Remarkable Mathematicians*, Cambridge University Press, Cambridge, 2002, pp. 271–8.

[43] For more details on this unusual phenomenon, see pages 706–7 in Carl B. Boyer and Uta C. Merzbach, *A History of Mathematics*, Second edition, John Wiley & Sons, New York, 1989 or pages 197–204 in Jeremy J. Gray, *The Hilbert Challenge*, Oxford University Press, 2000.

> A mathematician, like a painter or a poet, is a maker of patterns. If his patterns are more permanent than theirs, it is because they are made with *ideas*. A painter makes patterns with shapes and colours, a poet with words.... The mathematician's patterns, like the painter's or the poet's, must be *beautiful*; the ideas, like the colours or the words, must fit together in a harmonious way. Beauty is the first test: there is no permanent place in the world for ugly mathematics.[44]

Hardy is also known for his mathematical partnerships with two men—John Edensor Littlewood (1885–1977) and Srinivasa Aaiyangar Ramanujan (1887–1920). Prior to the twentieth century, a number of results were identified by the name of two individuals, such as the Bolzano–Weierstrass theorems, the Cauchy–Riemann equations, or the Euler–Maclaurin formula. This pairing of names should not suggest a collaboration on the part of these men, as they developed their results independent of contact with one another. But in the case of G. H. Hardy and John E. Littlewood, it was quite another matter. These two men knew each other and often discussed mathematical ideas. They had some unusual rules for working together, but a Hardy–Littlewood authorship led the way for a new level of collaborative writing that continues until the present day.

The story of Ramanujan[45] is one of the most unusual and mysterious in all of mathematics. In the year 1913, he was an unknown clerk working in a government office in Madras, India. During his sporadic schooling, he came across a mathematics text *Synopsis of Elementary Results in Pure and Applied Mathematics*, written by a British mathematician G.S. Carr, and learned much of his early mathematics from this book. He obtained addresses for some mathematicians in Britain and mailed them examples of mathematical results he had obtained, asking for an evaluation of their importance. One of these packets came to Hardy at Cambridge early in 1913. This strange collection of both known and new results in mathematical analysis puzzled Hardy at first, but he eventually came to believe that Ramanujan was an untrained genius. After a year of negotiations, arrangements were completed for Ramanujan to come to Cambridge and study with Hardy and Littlewood. During the next four years spent in Cambridge, Ramanujan obtained many new results and honors, including election as a fellow of the Royal Society. However, his health was precarious and he died only one year after returning to India in 1919. He wrote many of his results in notebooks, one of which caused excitement in the mathematical community when it was found in 1976 at Trinity College in Cambridge after being lost for many years. This led to a new interest in Ramanujan and resulted in his notebooks being printed in several volumes by Springer-Verlag.[46]

In 1936–37, Hardy spent a year at Princeton University, exchanging positions with the American mathematician Oswald Veblen (1880–1960). During this time, he met William Bullitt, a lawyer from Louisville, Kentucky, who had graduated from Princeton in 1894 with a mathematics major and a lifelong interest as an amateur in mathematical ideas. At

---

44 G.H. Hardy, *A Mathematician's Apology*, Cambridge University Press, Cambridge, 1969, pp. 84–5.

45 Robert Kanigel, *The Man Who Knew Infinity*, Charles Scribner's Sons, New York, 1991.

46 For example, see Bruce C. Berndt, *Ramanujan's Notebooks*, Part I, Springer-Verlag, New York, 1985.

the time he met Hardy, he was just beginning a 20-year search to collect rare mathematical books[47] written by the greatest mathematicians of all time. Bullitt asked Hardy, along with other outstanding mathematicians of the day, for their choice of the twenty-five greatest mathematicians. Hardy loved to evaluate and rank things, whether it be the skill of a cricket player, various proofs of the result that

$$\int_0^\infty \frac{\sin x}{x}\,dx = \frac{\pi}{2},$$

or the contribution of a great mathematician. Here are his choices, with the names of those who can be considered as contributors to analysis in capital letters. This underscores the fact that most great mathematicians worked on calculus and analysis during at least part of their career.

Hardy ranked the top three as follows: ARCHIMEDES, NEWTON, GAUSS.

The rest of his top ten in chronological order were: FERMAT, EULER, LAGRANGE, CAUCHY, ABEL, Galois, RIEMANN.

Again in chronological order, the final fifteen names were: Apollonius, Euclid, Eudoxus, DESCARTES, LEIBNIZ, BERNOULLI, LAPLACE, FOURIER, DIRICHLET, Lobachevsky, Poncelet, DEDEKIND, CANTOR, WEIERSTRASS, and Poincaré.

To conclude this final historical essay, we list three of the remaining unsolved problems in analysis.

1. Find a closed form expression for the value of the series

$$\sum_{n=1}^{\infty} \frac{1}{n^3}.$$

   We have such expressions for

$$\sum_{n=1}^{\infty} \frac{1}{n^k}$$

   for all even values of k, but none for the odd values of $k > 1$.

2. Decide whether Euler's constant γ[48] is rational or irrational. And if it is irrational, is it algebraic or transcendental?

---

[47]These materials may be viewed at the library of the University of Louisville in Kentucky. They include a 1482 copy of Euclid's *Elements* printed in Venice, a 1544 copy of Archimedes *Opera* printed in Basel, a 1543 Nuremburg copy of Copernicus' *De Revolutionibus*, a 1679 Toulouse copy of Fermat's *Varia Opera Mathematica*, and a 1687 first edition of Newton's *Principia*, being a presentation copy to Lord Halifax and containing some corrections in Newton's own writing.

[48]For an entire book on this number and its connections with other parts of mathematics, see Julian Havil, *Gamma: Exploring Euler's Constant*, Princeton University Press, Princeton, New Jersey, 2003.

3. The Riemann hypothesis,[49] which is part of complex analysis, is considered to be the most famous unsolved problem in mathematics, now that Fermat's Last Theorem was proved in the 1990s.

Section 2. New Looks at Calculus Content

Essay 7. Obtaining the Derivative Formulas

The derivative formulas are developed early in a calculus course, and it is unlikely that students remember the derivation details by the time they begin their study of analysis. It would be most helpful to read this essay at the beginning of a real analysis course, since it provides a review of this part of calculus, as well as giving a model of the logical development that characterizes work in real analysis.

There are two types of derivative formulas. Those of the first type apply to combinations of general functions and are given names such as the product rule, the quotient rule, and the chain rule. Those of the second type deal with the derivatives of specific functions and can be subdivided into four categories:

1. the power rule
2. the trigonometric functions, including their inverse functions
3. the logarithmic and exponential functions
4. functions defined by use of an integral

This essay is concerned with derivative formulas of the second type.

Category 1. Power rule

Although the power rule asserts that

$$\frac{d}{dx}x^r = rx^{r-1}$$

for all real numbers r, its proof is spread over four cases depending on the nature of the exponent r. The first case is when r is a natural number n. The definition of the derivative requires that we either use the binomial theorem to expand $(x+\Delta x)^n$ or else that we factor the expression $x^n - x_0^n$. When the exponent is a negative integer (i.e., $r = -n$), we write

$$\frac{d}{dx}(x^{-n}) = \frac{d}{dx}\left(\frac{1}{x^n}\right)$$

and use the result in the first case and the quotient rule. If the exponent is a rational number (i.e., $r = p/q$ with p, q integers), we observe that if $y = x^{p/q}$, then $y^q = x^p$, and use the result from the second case of the power rule involving integers, as well as the chain rule,

[49] Two very recent books dealing with the Riemann zeta function are John Derbyshire, *Prime Obsession*, Joseph Henry Press, Washington, D.C., 2003 and Marcus du Sautoy, *The Music of the Primes*, HarperCollins Publisher, New York, 2003.

to find dy/dx. Later in calculus, after studying the derivatives of logarithms, when $y = x^r$ with r irrational, we rewrite this as $\ln y = r \ln x$ before demonstrating once again that

$$\frac{dy}{dx} = rx^{r-1}.$$

In summary, then, we need algebraic techniques such as the binomial formula, factoring of polynomials, log and exponential identities, along with the quotient and chain rules to prove the various cases of the power rule.

There are alternate ways to obtain these formulas,[50] but since the quotient and chain rules are needed anyway, it is just as easy to use the approach described above.

Exercise 1. Prove that

$$\frac{d}{dx}x^n = nx^{n-1}$$

when n is a positive integer. Use two different derivative definitions, one using the

$$\lim_{\Delta x \to 0} \frac{(x + \Delta x)^n - x^n}{\Delta x}$$

form of the definition and the other using the

$$\lim_{x \to x_o} \frac{x^n - x_o^n}{x - x_o}$$

form of the definition.

Exercise 2. Provide details for proofs of the other three cases of the power rule.

Category 2. Trigonometric functions

We next turn our attention to the six trigonometric functions along with their six inverse functions. If we can find the derivative for any one of these twelve functions, then the other eleven formulas will follow from it. The usual first choice is $\sin x$. Assume that

$$\frac{d}{dx}(\sin x) = \cos x$$

has been established (we will prove this later in this essay). Then the formula for

$$\frac{d}{dx}(\csc x)$$

follows from the quotient rule, and the formula for

$$\frac{d}{dx}(\cos x)$$

---

[50] See p. 175 in [**29**].

follows from the identity

$$\cos x = \sin\left(\frac{\pi}{2} + x\right)$$

and the chain rule, along with other trig identities. The derivatives for $\tan x$, $\sec x$, and $\cot x$ then follow from the use of the quotient rule.

Exercise 3. Obtain the derivatives formulas for $\csc x$, $\cos x$, and $\tan x$ in the manner described in the paragraph above.

The derivatives for each of the six inverse functions are obtained by the common technique illustrated next for $\arcsin x$. Begin with $y = \arcsin x$, rewrite it as $x = \sin y$, and then obtain

$$y' = \frac{1}{\cos y}.$$

The final result is found using the fact that $x = \sin y$ to rewrite $\cos y$ as a function of x. Consider the graph of the function $y = \arcsin x$ to explain why the positive square root should be chosen in this case. This commentary supports the claim that the remaining eleven trigonometric derivatives can be found from the knowledge that

$$\frac{d}{dx}\sin x = \cos x.$$

Exercise 4. Provide the details for the proof outlined above that

$$\frac{d}{dx}\arcsin x = \frac{1}{\sqrt{1-x^2}}.$$

Exercise 5. In a similar way, provide details for obtaining the derivative formulas for $\arctan x$ and $\operatorname{arcsec} x$.

The initial formula (that $d/dx \sin x = \cos x$) is obtained by use of the definition of the derivative, some trigonometric identities, and the evaluation of the limit of an indeterminate form. The calculation

$$\begin{aligned}\frac{d}{dx}\sin x &= \lim_{\Delta x\to 0}\frac{\sin(x+\Delta x)-\sin x}{\Delta x}\\ &= \cos x\cdot\lim_{\Delta x\to 0}\frac{\sin\Delta x}{\Delta x} + \sin x\cdot\lim_{\Delta x\to 0}\frac{\cos\Delta x - 1}{\Delta x}\end{aligned}$$

shows that

$$\lim_{h\to 0}\frac{\sin h}{h} = 1 \quad\text{and}\quad \lim_{h\to 0}\frac{\cos h - 1}{h} = 0$$

must be verified, where we replace Δx by h to make the notation simpler. The second limit depends on the first, since

$$\frac{\cos h - 1}{h} = \frac{\cos h - 1}{h}\cdot\frac{\cos h + 1}{\cos h + 1} = \frac{-\sin^2 h}{h(\cos h + 1)} = \frac{-\sin h}{\cos h + 1}\cdot\frac{\sin h}{h}.$$

The evaluation of the final key limit that

$$\lim_{h\to 0} \frac{\sin h}{h} = 1$$

depends on some basic results about trigonometry and the unit circle, a comparision of three areas, and the squeeze theorem for limits.[51]

Exercise 6. Write a careful proof that

$$\frac{d}{dx} \sin x = \cos x,$$

including all necessary details. The outline for such a proof is given above.

Exercise 7. In Exercise 1, we proved the power rule by two different definitions. Explain why the second form of the definition does not work for $\sin x$.

Category 3. Exponential and logarithmic functions

We see here a similar situation as in Category 2. If we can find the derivative for any one of the functions in Category 3, such as

$$\frac{d}{dx}(2^x) \quad \text{or} \quad \frac{d}{dx}(\log_{10} x),$$

then the derivative formulas of all the others will follow. As a first choice, we begin with $y = 2^x$ and show by use of the derivative definition that $y' = C \cdot 2^x$, where

$$C = \lim_{h\to 0} \frac{2^h - 1}{h}.$$

From this, we find

$$\frac{d}{dx} \log_2 x = \frac{1}{Cx}.$$

Since

$$\log_b x = \frac{\log_2 x}{\log_2 b},$$

we can show that

$$\frac{d}{dx} \log_b x = \frac{1}{C \log_2 b} \frac{1}{x}$$

for any $b > 0$ and then that

$$\frac{d}{dx} b^x = (C \log_2 b)\, b^x.$$

[51]This evaluation was outlined in Problem 4 of the Calculus Review problems in Part I.

Thus we have derivative formulas for the general exponential and logarithmic functions expressed in terms of C and $\log_2 b$. Although this may seem a reasonable way to proceed, it is not the standard approach found in a calculus textbook.

A second choice (which is the one usually chosen) is to say that

$$\int_1^x \frac{1}{t}\,dt$$

is the natural logarithm function, (i.e., $\ln x = \log_e x$, where $e \approx 2.71828$), leading to the result that

$$\frac{d}{dx}\ln x = \frac{1}{x}$$

(see Exercise 11). Once this is done, it is straightforward to find derivative formulas for the general logarithmic and exponential functions, as in the example above involving $y = 2^x$. But this statement that

$$\int_1^x \frac{1}{t}\,dt$$

is the function $\log_e x$ requires a large leap of faith, and it is only reasonable to expect some evidence be given to support such a claim. The next part of this essay describes how this might be done.

Exercise 8. If $f(x) = 2^x$, prove that $f'(x) = f'(0)\,f(x)$.

Exercise 9. Assuming that

$$\frac{d}{dx}2^x = C \cdot 2^x,$$

show that

$$\frac{d}{dx}b^x = (C \log_2 b)b^x.$$

Exercise 10. Find a reasonable approximation for

$$C = \lim_{h\to 0}\frac{2^h - 1}{h}.$$

Exercise 11. Use the definition of the derivative and the squeeze theorem to prove that

$$\frac{d}{dx}\int_1^x \frac{1}{t}\,dt = \frac{1}{x} \quad \text{for } x > 0.$$

Evidence that $G(x) = \int_1^x \frac{1}{t}\,dt$ is a logarithmic function. We begin with a list of properties that we typically associate with a general logarithmic function, denoted by $F(x) = \log_b x$.

1. $F(1) = 0$.
2. The domain of $F(x)$ is all $x > 0$.
3. $F(x)$ is an increasing function that is concave down.
4. $\lim_{x\to 0^+} F(x) = -\infty$ and $\lim_{x\to\infty} F(x) = +\infty$.
5. $F(xy) = F(x) + F(y)$.
6. $F(x^r) = r\,F(x)$.
7. $F(b) = 1$.

The next step is to verify that the function

$$G(x) = \int_1^x \frac{1}{t}\,dt$$

satisfies these seven properties. This increases our confidence that $G(x)$ is a logarithm function.

Theorem 1. $G(xy) = G(x) + G(y)$.

Proof At first glance, it looks intimidating to try to prove that

$$\int_1^{xy} \frac{1}{t}\,dt = \int_1^x \frac{1}{t}\,dt + \int_1^y \frac{1}{t}\,dt.$$

We simplify our work by choosing any values x and a in the domain of G, considering x as a variable and a as a constant. We already know that

$$G'(x) = \frac{d}{dx}\int_1^x \frac{1}{t}\,dt = \frac{1}{x},$$

and so by use of the chain rule,

$$\frac{d}{dx}G(ax) = \frac{d}{dx}\int_1^{ax} \frac{1}{t}\,dt = \frac{1}{ax}\cdot a = \frac{1}{x} = \frac{d}{dx}G(x).$$

Since two functions have identical derivatives, it follows by a corollary to the mean value theorem that there is some constant k so that $G(ax) = G(x) + k$ for all x. By setting $x = 1$, we get $G(a) = G(1) + k$ or $k = G(a)$. Hence we have $G(ax) = G(x) + G(a)$. Since x and a are arbitrary values in the domain of G, the proof is complete. ■

Exercise 12. Verify that $G(x)$ satisfies properties 1–4 listed above for a logarithm function.

Exercise 13. Prove Property 6 which states that $G(x^r) = r\,G(x)$. Hint: Adapt the method used to prove Theorem 1.

Exercise 14. Give an example of a non-logarithm function that satisfies Properties 1–4.

In order to find the base b so that $G(x) = \log_b x$, we observe that $G(b) = 1$ must hold, so we need to solve

$$\int_1^b \frac{1}{t}\,dt = 1 \quad \text{for } b.$$

The trapezoid rule gives

$$\int_1^2 \frac{1}{t}\,dt \approx .7 \quad \text{and} \quad \int_1^3 \frac{1}{t}\,dt \approx 1.1,$$

so we are looking for a value of b between 2 and 3, but closer to 3. By increasing the accuracy of our work, we can find values of b close to the familiar value for e.

Here is a third and final choice for the beginning derivative formula in Category 3, namely to find $F'(x)$ when $F(x) = \log_b x$. Assuming that

$$\lim_{n\to\infty}\left(1+\frac{1}{n}\right)^n = e$$

is already known, we observe that by substituting $x/\Delta x$ for n,

$$e = \lim_{n\to\infty}\left(1+\frac{1}{n}\right)^n = \lim_{\Delta x\to 0^+}\left(1+\frac{\Delta x}{x}\right)^{x/\Delta x}.$$

It also follows from logarithmic properties that

$$\begin{aligned}\frac{\log_b(x+\Delta x)-\log_b x}{\Delta x} &= \frac{1}{\Delta x}\log_b\left(\frac{x+\Delta x}{x}\right)\\ &= \frac{x}{\Delta x}\,\frac{1}{x}\log_b\left(1+\frac{\Delta x}{x}\right)\\ &= \frac{1}{x}\log_b\left(1+\frac{\Delta x}{x}\right)^{x/\Delta x}.\end{aligned}$$

By substituting these results into the derivative definition, we have

$$\begin{aligned}F'(x) = \frac{d}{dx}\log_b x &= \lim_{\Delta x\to 0}\frac{\log_b(x+\Delta x)-\log_b x}{\Delta x}\\ &= \lim_{\Delta x\to 0}\frac{1}{x}\log_b\left(1+\frac{\Delta x}{x}\right)^{x/\Delta x}\\ &= \frac{1}{x}\log_b\left(\lim_{\Delta x\to 0^+}\left(1+\frac{\Delta x}{x}\right)^{x/\Delta x}\right)\\ &= \frac{1}{x}\log_b e = \frac{1}{x\ln b}.\end{aligned}$$

This result adds evidence to the assumption that

$$G(x) = \int_1^x \frac{1}{t}\,dt$$

is a logarithm function since $d/dx \log_b x$ contains a factor of $1/x$, as does $G'(x)$. The commentary under Category 3 is not found in a calculus text (or in most analysis texts). It describes three different ways to obtain the beginning derivative formula for the collection of exponential and logarithmic functions.

Exercise 15. After reviewing the above commentary under Category 3, describe the advantages and disadvantages in obtaining each of the following derivative formulas as the first one in this category.

1. $\dfrac{d}{dx}2^x$
2. $\dfrac{d}{dx}\displaystyle\int_1^x \frac{1}{t}\,dt$
3. $\dfrac{d}{dx}\log_b x$

Category 4. Functions defined using an integral

We have already encountered $\int_1^x 1/t\,dt$, which is an important example of this category of function. In fact, part of the fundamental theorem of calculus is a statement of how to find the derivative for such a function, namely that if $F(x) = \int_a^x f(t)\,dt$, then $F'(x) = f(x)$ for any value x for which $f(x)$ is continuous.

A second type of function defined using an integral is

$$G(x) = \int_a^b g(x,t)\,dt,$$

where $g(x,t)$ is a function of two variables. By use of the definition of the derivative and some basic integral properties, it can be shown that

$$G'(x) = \int_a^b g_x(x,t)\,dt,$$

where $g_x(x,t)$ represents the partial derivative of $g(x,t)$ with respect to x. Probably the most familiar example of such a function is the gamma function defined as

$$\Gamma(x) = \int_0^\infty t^{x-1}e^{-t}\,dt$$

(see Problem 29).

In summary, we list below the first step in the derivations of seven different derivative formulas. In each case, an appropriate algebraic identity is needed as the next step to simplify these expressions.

Exercise 16. Write the next algebraic step or two in the proof of the derivative formulas for each of the seven functions below. This should help you to see a common structure in the proofs of all these derivative formulas.

$$\frac{d}{dx}(x^n) = \lim_{\Delta x\to 0} \frac{(x+\Delta x)^n - x^n}{\Delta x} \tag{1}$$

$$\frac{d}{dx}\sqrt{x} = \lim_{\Delta x\to 0} \frac{\sqrt{x+\Delta x} - \sqrt{x}}{\Delta x} \tag{2}$$

$$\frac{d}{dx}(\sin x) = \lim_{\Delta x \to 0} \frac{\sin(x + \Delta x) - \sin x}{\Delta x} \tag{3}$$

$$\frac{d}{dx}(2^x) = \lim_{\Delta x \to 0} \frac{2^{x+\Delta x} - 2^x}{\Delta x} \tag{4}$$

$$\frac{d}{dx}(\log_b x) = \lim_{\Delta x \to 0} \frac{\log_b(x + \Delta x) - \log_b x}{\Delta x} \tag{5}$$

$$\frac{d}{dx}\left(\int_1^x \frac{1}{t}\,dt\right) = \lim_{\Delta x \to 0} \frac{\int_1^{x+\Delta x} \frac{1}{t}\,dt - \int_1^x \frac{1}{t}\,dt}{\Delta x} \tag{6}$$

$$\frac{d}{dx}(f \cdot g)(x) = \lim_{\Delta x \to 0} \frac{f(x + \Delta x)g(x + \Delta x) - f(x)g(x)}{\Delta x} \tag{7}$$

Essay 8. Tests for Convergence of Series

The best time to read this essay is after completing the introductory unit in analysis on sequences, and learning about results such as that a bounded and monotone sequence converges and that Cauchy sequences converge. Beginning with the definition of series convergence, and continuing with proofs of the tests for convergence, material in a unit on infinite series is based on results from the unit on sequences. The purpose of this essay is to emphasize this fact in a way that is not done in a calculus text, where there seems to be no unifying idea among the convergence tests. Hopefully this will provide a better understanding of this topic.

Definition 1. The series $\sum_{i=1}^{\infty} a_i$ converges to the real number L if the sequence of partial sums $\{S_n\}$ converges to L, where $S_n = \sum_{i=1}^{n} a_i$.

Section 1. Relation between the sequences $\{a_n\}$ and $\{S_n\}$

In the early 1800s, it was believed that if $\lim_{n\to\infty} a_n = 0$, then the series $\sum_{n=1}^{\infty} a_n$ must converge because one was adding smaller and smaller terms to the sum. We now know this is not true, because the harmonic series

$$\sum_{n=1}^{\infty} \frac{1}{n}$$

diverges to ∞ even though

$$\lim_{n\to\infty} \frac{1}{n} = 0.$$

However, the converse statement is true, as we prove in Theorem 1. Even the stronger condition that $\lim_{n\to\infty} na_n = 0$ is not enough to prove convergence, as the divergent series

$$\sum_{n=1}^{\infty} \frac{1}{n \ln n}$$

demonstrates. But it can be proven by the limit form of the comparison test that the series $\sum_{n=1}^{\infty} a_n$ converges if $\lim_{n\to\infty} n^k a_n = 0$ for some $k > 1$.

Theorem 1. *If the series $\sum_{n=1}^{\infty} a_n$ converges to L, then $\lim_{n\to\infty} a_n = 0$.*

Proof Since $a_n = S_n - S_{n-1}$, it follows that

$$\lim_{n\to\infty} a_n = \lim_{n\to\infty} S_n - \lim_{n\to\infty} S_{n-1} = L - L = 0. \qquad \blacksquare$$

The contrapositive statement of Theorem 1 is called the nth term test and states that if $\lim_{n\to\infty} a_n \neq 0$, then the series $\sum_{n=1}^{\infty} a_n$ diverges.

Section 2. Two important examples of convergent series

Example 1. The geometric series $\sum_{n=0}^{\infty} r^n$ converges to the value $1/(1-r)$ if $|r| < 1$.

Solution. If we let

$$S_n = 1 + r + r^2 + \cdots + r^{n-1},$$

then by cancellation of terms (see Problem 3), it follows that $S_n - rS_n = 1 - r^n$. From this, we have

$$S_n = \frac{1-r^n}{1-r}, \quad \text{so} \quad \lim_{n\to\infty} S_n$$

exists if $|r| < 1$ and has the value $1/(1-r)$. $\blacksquare$

Example 2. The p-series

$$\sum_{n=1}^{\infty} \frac{1}{n^p}$$

converges if $p > 1$.

Solution. Since

$$f(x) = \frac{1}{x^p}$$

is decreasing, it follows that

$$\frac{1}{n^p} \leq \frac{1}{x^p} \quad \text{for all} \quad x \leq n,$$

and so

$$\frac{1}{n^p} = \int_{n-1}^{n} \frac{1}{n^p}\,dx \le \int_{n-1}^{n} \frac{1}{x^p}\,dx.$$

Therefore,

$$\begin{aligned}
1 + \frac{1}{2^p} + \frac{1}{3^p} + \cdots + \frac{1}{n^p} &\le 1 + \int_1^2 \frac{1}{x^p}\,dx + \int_2^3 \frac{1}{x^p}\,dx + \cdots + \int_{n-1}^{n} \frac{1}{x^p}\,dx \\
&= 1 + \int_1^n \frac{1}{x^p}\,dx \\
&= 1 + \left.\frac{x^{1-p}}{1-p}\right|_1^n \\
&= 1 - \frac{1}{1-p} - \frac{1}{(p-1)n^{p-1}} \\
&< 1 - \frac{1}{1-p} = \frac{p}{p-1}.
\end{aligned}$$

Since S_n is bounded above, the series

$$\sum_{n=1}^{\infty} \frac{1}{n^p}$$

must converge, and to a value less than $p/(p-1)$. This argument is generalized later to prove the integral test. ■

Section 3. Proofs of convergence tests for series with positive terms

If all the terms a_n in the series $\sum_{n=1}^{\infty} a_n$ are positive, then the sequence of partial sums $\{S_n\}$ is monotone increasing. So this sequence (and hence the series also) will converge if and only if it is bounded above. Therefore, the key idea in the proofs of the following familiar tests—the comparison test, the integral test, and the ratio test—is based simply on some technique for finding an upper bound for S_n. Notice this common feature in the proofs of Theorems 2–7.

Theorem 2. (Direct comparison test) *If $\sum_{n=1}^{\infty} c_n$ converges and $0 < a_n \le c_n$ for all $n > N$, then $\sum_{n=1}^{\infty} a_n$ converges also.*

Proof The general term of the sequence of partial sums

$$C_n = \sum_{i=1}^{n} c_i$$

is bounded above, say by C. If

$$A_n = \sum_{i=1}^{n} a_i,$$

then for $n > N$,

$$A_n - A_N \leq C_n - C_N,$$

or

$$A_n \leq C_n - C_N + A_N < C - C_N + A_N.$$

Since $\{A_n\}$ is bounded above, the series $\sum_{n=1}^{\infty} a_n$ must converge. ∎

Example 3. The series

$$\sum_{n=0}^{\infty} \frac{1}{n!}$$

is important since it converges to the value e.

Solution. We know the geometric series

$$\sum_{n=0}^{\infty} \frac{1}{2^n}$$

converges to 2, and that

$$\frac{1}{n!} \leq \frac{1}{2^n}$$

for all $n > 3$. Therefore the series

$$\sum_{n=0}^{\infty} \frac{1}{n!}$$

converges by Theorem 2 and to a value less than

$$2 - C_4 + A_4 = 2 - \left(1 + \tfrac{1}{2} + \tfrac{1}{4} + \tfrac{1}{8}\right) + \left(1 + 1 + \tfrac{1}{2} + \tfrac{1}{6}\right) = 2\tfrac{19}{24} \approx 2.791.$$ ∎

Exercise 1. Repeat the work in Example 3 to find an upper bound for

$$\sum_{n=0}^{\infty} \frac{1}{n!}$$

when comparing it to the series

$$\sum_{n=0}^{\infty} \frac{1}{3^n}.$$

Lemma 1. *If $\sum_{n=1}^{\infty} c_n$ converges, then $\sum_{n=1}^{\infty} k\, c_n$ converges for any $k > 0$.*

Proof Let

$$C_n = c_1 + c_2 + \cdots + c_n.$$

The sequence $\{C_n\}$ converges if and only if $\{k\,C_n\}$ converges. Hence the series $\sum_{n=1}^{\infty} c_n$ converges if and only if $\sum_{n=1}^{\infty} k\,c_n$ converges. ■

Theorem 3. (Limit form of the comparision test) *If $\sum_{n=1}^{\infty} c_n$ converges and*

$$\lim_{n\to\infty} \frac{a_n}{c_n} = L \neq \infty,$$

then $\sum_{n=1}^{\infty} a_n$ converges also.

Proof By definition of a limit, there is some N so that

$$\frac{a_n}{c_n} < L + 1 \quad \text{for all } n > N.$$

But $a_n < (L+1)\,c_n$ for $n > N$ implies that $\sum_{n=1}^{\infty} a_n$ converges by the direct comparison test and Lemma 1. ■

Theorem 4. (Ratio form of the comparison test) *If*

$$\frac{a_{n+1}}{a_n} \leq \frac{c_{n+1}}{c_n}$$

for all n and if $\sum_{n=1}^{\infty} c_n$ converges, then $\sum_{n=1}^{\infty} a_n$ converges also.

Proof

$$\begin{aligned} a_n &= a_1 \cdot \frac{a_2}{a_1} \cdot \frac{a_3}{a_2} \cdots \frac{a_n}{a_{n-1}} \\ &\leq a_1 \cdot \frac{c_2}{c_1} \cdot \frac{c_3}{c_2} \cdots \frac{c_n}{c_{n-1}} \\ &= \frac{a_1}{c_1} \cdot c_n \end{aligned}$$

so $\sum_{n=1}^{\infty} a_n$ converges by use of the direct comparison test and Lemma 1. Note that the theorem is also true if

$$\frac{a_{n+1}}{a_n} \leq \frac{c_{n+1}}{c_n}$$

holds only for all $n >$ some N. ■

Theorem 5. (Ratio test) *If*

$$\lim_{n\to\infty} \frac{a_{n+1}}{a_n} = L < 1,$$

then $\sum_{n=1}^{\infty} a_n$ converges.

Proof Since

$$\lim_{n\to\infty} \frac{a_{n+1}}{a_n} = L < 1,$$

there is some $r < 1$ and some N so that

$$\frac{a_{n+1}}{a_n} < r \quad \text{for all } n > N.$$

Since the geometric series

$$\sum_{n=1}^{\infty} c_n = \sum_{n=1}^{\infty} r^n$$

converges and

$$\frac{c_{n+1}}{c_n} = \frac{r^{n+1}}{r^n} = r,$$

it follows that $\sum_{n=1}^{\infty} a_n$ converges by the ratio form of the comparison test. ■

Example 4. For the series

$$\sum_{n=0}^{\infty} \frac{1}{n!},$$

we have

$$\frac{a_{n+1}}{a_n} = \frac{1}{n+1},$$

and this is less than $r = \frac{1}{4}$ when $n > 3 = N$. So an upper bound for this series is

$$\frac{a_3}{1-\frac{1}{4}} + A_2 = \frac{\frac{1}{3!}}{\frac{3}{4}} + \left(1 + 1 + \frac{1}{2}\right) = 2\frac{13}{18} \approx 2.7\overline{2}.$$ ■

Exercise 2. For the same series

$$\sum_{n=0}^{\infty} \frac{1}{n!}$$

as in Example 4, use $r = \frac{1}{6}$ to find the upper bound of

$$2\tfrac{69}{96} \approx 2.71875.$$

Theorem 6. (Integral test) *Let $f(x)$ be a continuous and decreasing function so that the improper integral $\int_1^{\infty} f(x)\,dx$ converges to L, and let $a_n = f(n)$. Then the series $\sum_{n=1}^{\infty} a_n$ converges also and to a value less than $a_1 + L$.*

Proof Since $f(x)$ is decreasing, it follows that $a_n \le f(x)$ for $x \in [n-1, n]$ and so $a_n \le \int_{n-1}^{n} f(x)\,dx$ for $n > 1$. It follows that

$$\begin{aligned} S_n &= a_1 + a_2 + a_3 + \cdots + a_n \\ &\le a_1 + \int_1^2 f(x)\,dx + \int_2^3 f(x)\,dx + \cdots + \int_{n-1}^{n} f(x)\,dx \\ &= a_1 + \int_1^n f(x)\,dx \\ &\le a_1 + \int_1^\infty f(x)\,dx = a_1 + L \end{aligned}$$

and so

$$\sum_{n=1}^{\infty} a_n = \lim_{n\to\infty} S_n \le a_1 + L.$$

■

A decreasing sequence of upper bounds can be obtained by generalizing the above argument as follows.

$$\begin{aligned} S_n &= a_1 + a_2 + a_3 + \cdots + a_n \\ &< a_1 + a_2 + \cdots + a_k + \int_k^{k+1} f(x)\,dx + \int_{k+1}^{k+2} f(x)\,dx + \cdots + \int_{n-1}^{n} f(x)\,dx \\ &= S_k + \int_k^n f(x)\,dx \\ &< S_k + \int_k^\infty f(x)\,dx \end{aligned}$$

and so

$$\sum_{n=1}^{\infty} a_n = \lim_{n\to\infty} S_n \le S_k + \int_k^\infty f(x)\,dx.$$

Exercise 3. In a similar way, show that

$$S_{k-1} + \int_k^\infty f(x)\,dx \le \sum_{n=1}^{\infty} a_n.$$

The above results show that $S_k + \int_k^\infty f(x)\,dx$ is an approximation for the value of the series $\sum_{n=1}^{\infty} a_n$ with an error less than a_k. The meaning of these inequalities can be better understood by considering the terms involved as areas.

Theorem 7. (Root test)[52] *If*

$$\lim_{n\to\infty} \sqrt[n]{a_n} = L < 1, \quad \textit{then} \quad \sum_{n=1}^{\infty} a_n$$

converges.

[52] See Problem 10 for an extension and comparison of the ratio and root tests.

Proof We choose a value r so that $L < r < 1$. Then there is some N so that

$$\sqrt[n]{a_n} < r \quad \text{for all } n \geq N.$$

Since this is equivalent to $a_n < r^n$ for all $n \geq N$ and since $\sum_{n=1}^{\infty} r^n$ is a convergent geometric series, it follows that $\sum_{n=1}^{\infty} a_n$ converges by the direct comparison test. ■

Section 4. Tests for series with mixed signs

Theorem 8. (Alternating series test)[53] *If $\{a_n\}$ is such that $0 < a_{n+1} \leq a_n$ for all n and $\lim_{n\to\infty} a_n = 0$, then the series of alternating signs*

$$\sum_{n=1}^{\infty} (-1)^{n+1} a_n$$

converges, and to a value L so that $S_{2n} < L < S_{2n+1}$.

Proof The hypotheses for this test guarantee that the subsequence $\{S_{2n}\}$ is an increasing sequence that is bounded above, while the subsequence $\{S_{2n-1}\}$ is a decreasing sequence that is bounded below. The fact that S_{2n} and S_{2n-1} can be written in the following way, noting the terms in parentheses are all non-negative, helps to verify these assertions.

$$S_{2n} = (a_1 - a_2) + (a_3 - a_4) + \cdots + (a_{2n-1} - a_{2n})$$

$$S_{2n} = a_1 - (a_2 - a_3) - \cdots - (a_{2n-2} - a_{2n-1}) - a_{2n} < a_1$$

$$S_{2n-1} = a_1 - (a_2 - a_3) - \cdots - (a_{2n-2} - a_{2n-1})$$

So both subsequences converge. Assume

$$\lim_{n\to\infty} S_{2n} = L \quad \text{and} \quad \lim_{n\to\infty} S_{2n-1} = M.$$

Since

$$a_{2n} = S_{2n} - S_{2n-1} \quad \text{and} \quad \lim_{n\to\infty} a_n = 0,$$

it follows that

$$\lim_{n\to\infty} a_{2n} = \lim_{n\to\infty} S_{2n} - \lim_{n\to\infty} S_{2n-1} = L - M = 0.$$

Since $L = M$, both the sequence $\{S_n\}$ and the series

$$\sum_{n=1}^{\infty} (-1)^{n-1} a_n$$

converge to L. ■

S_{2n} approximates L with an error less than a_{2n+1}.[54] For an alternating series, when the desired error for an approximation is given, this tells us how many terms need to be summed to obtain such an approximation.

---

[53] This test is generalized in Problem 13.

[54] The midpoint value $\frac{1}{2}(S_{2n} + S_{2n+1})$ approximates L with an error less than $\frac{1}{2}a_{2n+1}$.

Theorem 9. (Absolute convergence test) *If the series* $\sum_{n=1}^{\infty} |a_n|$ *converges, then the series of mixed signs* $\sum_{n=1}^{\infty} a_n$ *converges also.*

Proof The concept of a Cauchy sequence is the key to this proof.[55] We first observe that the following five statements are all equivalent.

$$\text{The series } \sum a_n \text{ converges.} \tag{1}$$

$$\text{The sequence } \{S_n\} \text{ where } S_n = a_1 + a_2 + \cdots + a_n \text{ converges.} \tag{2}$$

$$\text{The sequence } \{S_n\} \text{ is a Cauchy sequence.} \tag{3}$$

$$\forall \epsilon > 0,\ \exists N_\epsilon \text{ so that for all } m > k > N_\epsilon,\ |S_m - S_k| < \epsilon. \tag{4}$$

$$\forall \epsilon > 0, \exists N_\epsilon \text{ so that for all } m > k > N_\epsilon,\ |a_{k+1} + a_{k+2} + \cdots + a_m| < \epsilon. \tag{5}$$

These statements tell us that $\sum_{n=1}^{\infty} |a_n|$ converges if and only if $\forall \epsilon > 0, \exists N_\epsilon$ so that for all

$$m > k > N_\epsilon,\ ||a_{k+1}| + |a_{k+2}| + \cdots + |a_m|| < \epsilon.$$

The fact that

$$|a_{k+1} + a_{k+2} + \cdots + a_m| \leq |a_{k+1}| + |a_{k+2}| + \cdots + |a_m|$$

is enough to complete the proof. ∎

Summary. Notice that in the proof of every one of the convergence tests given above, our goal is to prove that the sequence of partial sums $\{S_n\}$ converges. When all the terms a_n are positive, as they are for Theorems 2–7, $\{S_n\}$ is an increasing sequence, so we only need to find an upper bound to prove convergence. For an alternating series, we turn to subsequences and show that $\{S_{2n}\}$ and $\{S_{2n-1}\}$ are monotone and bounded, and hence convergent. The proof of Theorem 8 is completed by proving that both subsequences converge to the same value. For the absolute convergence test (Theorem 9), the key is to use a Cauchy sequence and the fact that all Cauchy sequences are convergent, and vice-versa.

Section 3. General Topics for Analysis

Essay 9. Proof Techniques in Analysis

It is best to read this essay after you have taken several weeks of an analysis class and have some familiarity with the theorems discussed there. The purpose of this essay is to provide basic comments about proof techniques, with all examples drawn from the subject matter of real analysis.

Some twenty-four centuries ago, Euclid set down the standard of mathematical reasoning which also characterizes our present day point-of-view. In his book, the *Elements*,

[55] The Weierstrass M-test is a second theorem whose proof needs this concept. This test is used to verify uniform convergence for a series of functions.

Euclid presented a model for the axiomatic method, when he began with a list of definitions and axioms and then developed the proofs of several hundred theorems, arranging everything in a logical order. Though there were some weaknesses in his work, it provides a masterful pattern for us to follow today.

The concept of proof is an essential part of mathematics. Our understanding of what constitutes a proof has varied greatly during the historical development of mathematics. There is now essentially universal agreement that a correct proof needs a logical development with every step supported by an axiom, definition, or a previously proven result. Although this belief has characterized mathematics during the past century, a more intuitive approach was in vogue prior to this time.

A variety of proof forms

We begin by considering a variety of proof forms and techniques, such as a direct proof, a contrapositive proof, an indirect proof, the law of *modus ponens*, and the principle of mathematical induction.

A typical theorem is of the form $p \Rightarrow q$. The statement p is the collection of hypotheses that are assumed, and the statement q is the desired conclusion. Our first attempt is to find a *direct proof*, which may consist of several separate implications such as $p \Rightarrow s$ and $s \Rightarrow t$ and $t \Rightarrow q$. Then $p \Rightarrow q$ follows from the transitive property of $\Rightarrow$.

It may be that the hypotheses in p are not enough to obtain q by any direct proof attempt. We then consider the statement $\neg q$, which is the negation of q. If we can find a direct proof that $\neg q \Rightarrow \neg p$, we claim that $p \Rightarrow q$ is also true. This is because these are *contrapositive* statements and therefore logically equivalent. The proof of the nth term test for series is an example of this idea.

If we fail in an attempt to find a direct proof that $p \Rightarrow q$ or that $\neg q \Rightarrow \neg p$, we assume p and $\neg q$ together. If we are able to find some statement t so that t and $\neg t$ are both consequences of p and $\neg q$, we claim to have found a contradiction, and assert that $p \Rightarrow q$ is true. This is considered an *indirect proof*, since we have not actually proven that q is true, just that $\neg q$ cannot be true. This is called the law of the excluded middle, which has not always been considered a valid form of reasoning. However, most mathematicians today accept an indirect proof as valid and use it whenever a direct proof cannot be found. Many standard mathematical results, such as that $\sqrt{2}$ is irrational, can only be proven by the indirect method. These would not be accepted as valid by those who hold to a strict view of intuitionism.

The result that a product of convergent sequences is also convergent, or that a continuous function is integrable are examples of results we can verify by a direct proof. The two results that a continuous function defined on a compact domain is bounded and also uniformly continuous are examples of theorems that are usually proven by the indirect method. The majority of analysis proofs are of the direct form.

Most mathematical proofs are written in essay form, rather than the step-reason format often used in a high school geometry course. The following proof is written in step-reason form to help you better understand the logical structure of this proof. The goal is to provide a formal proof that a continuous function defined on a closed interval is bounded.

Example 1.

p: f is continuous on $[a, b]$
q: f is bounded on $[a, b]$

Proof

$\neg q$: f is not bounded on $[a, b]$
r: $\forall n, \exists$ some $x_n \in [a, b]$ so that $|f(x_n)| > n$ and also $|f(x_n)| > |f(x_{n-1})|$
s: $\{x_n\}$ is a bounded sequence
t: there is a subsequence $\{x_{n_k}\}$ of $\{x_n\}$ and some x_o so $\lim_{k\to\infty} x_{n_k} = x_o$
u: $x_o \in [a, b]$
v: $\lim_{k\to\infty} f(x_{n_k}) = +\infty$
$\neg v$: $\lim_{k\to\infty} f(x_{n_k}) = f(x_o)$ ■

Each statement is a direct consequence of a previous statement or statements. For example, $s \Rightarrow t$ because of the Bolzano–Weierstrass theorem, whereas $r \wedge t \Rightarrow v$. Why does $t \Rightarrow u$? Which statements imply $\neg v$?

Because the statements v and $\neg v$ have been proven from the assumptions of p and $\neg q$, this contradiction completes an indirect proof of the desired result that $p \Rightarrow q$. Notice that the above proof can easily be changed into a contrapositive proof form by replacing the statement $\neg v$ above by the statement $\neg p$ as follows.

$\neg p$: f is not continuous at x_o and therefore not continuous on $[a, b]$

The law of *modus ponens* refers to the logical statement $\{p \wedge [p \Rightarrow q]\} \Rightarrow q$. It is one of the two rules for proving logical statements that Bertrand Russell and Alfred North Whitehead assumed in their multi-volumed book *Principia Mathematica*. This law is used to prove new theorems from old (or previously proven) theorems, and might be more accurately represented by $\{p_{\text{new}} \wedge [p_{\text{old}} \Rightarrow q_{\text{old}}]\} \Rightarrow q_{\text{new}}$.

Example 2. We illustrate this law by showing the (new) theorem, which asserts an increasing sequence that is bounded above must converge, can be proved by using the least upper bound axiom (as the old result).

p_{new}: The sequence $\{x_n\}$ is increasing and bounded above.
q_{new}: The sequence $\{x_n\}$ converges.
p_{old}: The set S is non-empty and bounded above.
q_{old}: The set S has a least upper bound (i.e., sup S exists).

The proof begins by choosing a sequence $\{x_n\}$ that is increasing and bounded above. In order to use the least upper bound axiom, we must identify a set S that is non-empty and has an upper bound. In this case, it is fairly obvious that our only realistic choice is to define $S = \{x_n : n \in N\}$. We then assert that sup S exists by using the least upper bound axiom. The proof is completed by showing $\{x_n\}$ converges to sup S. ■

We just observed that the law of modus ponens is an axiom of logical reasoning, which means we accept it as valid without a proof. In a similar way, the principle of *mathematical induction* is an axiom presented by Giuseppe Peano (1858–1932) as one of the postulates he and Richard Dedekind used to define the set of natural numbers.[56] Its statement is familiar to most mathematics students: If $P(n)$ is a statement defined for each natural number n with the two properties that $P(n+1)$ is true whenever $P(n)$ is true and $P(k)$ is true, then $P(n)$ is true for all $n \geq k$.

It can be used to verify statements such as

$$P(n) : 1 + 2 + 3 + \cdots + n = \frac{n(n+1)}{2} \quad \text{for all } n \quad \text{or} \quad P(n) : n! > 2^n$$

for $n \geq 4$. In analysis, we use this principle to prove the Bolzano–Weierstrass theorems for sets and sequences, when we inductively define a sequence of intervals with certain properties.

Having completed our initial discussion of what constitutes a valid proof, we next offer some practical ideas of how you can develop your ability to understand existing proofs and to also discover proofs on your own.

Suggestions for improving your skill in understanding proofs

1. Memorize basic definitions and statements of theorems.

   How can you hope to understand a proof of the Bolzano–Weierstrass theorem for sets without complete knowledge of the hypothesis and conclusion of this theorem, along with the meaning of the term "cluster point"? Don't just read the definitions and theorems several times. Speak them out loud, write them down, and have a dialogue with another person about them. The list of definitions and main theorems in any unit is relatively short, so this is a reasonable task to undertake. College students routinely do more difficult things than learning the statement and proof for the fundamental theorem of calculus or the Bolzano–Weierstrass theorem, such as playing ninety minutes in a soccer match or performing a lengthy musical composition that has been memorized.

2. Put definitions and theorems into collections with common properties.

   For example, consider all the definitions that involve the word "point" such as a limit point of a sequence, a cluster point for a set, a point of discontinuity for a function, and interior, exterior, or boundary points for a set. Another idea is to compare all the theorems about sequences that use the words *convergent* and *bounded* in their statement, such as:

   (a) A *convergent* sequence is *bounded*.

   (b) A *bounded* and monotone sequence is *convergent*.

   (c) A *bounded* sequence has a *convergent* subsequence.

[56] See Raymond L. Wilder, *Introduction to the Foundations of Mathematics*, John Wiley & Sons, New York, 1965, p. 158.

3. Use a flow chart diagram for theorems.

   This practice enables you to see clearly the sequential order of theorems. Such a diagram also gives insight into the content of a given proof, for it will likely involve the theorem to its left along with the law of *modus ponens*. Here are three examples.[57]

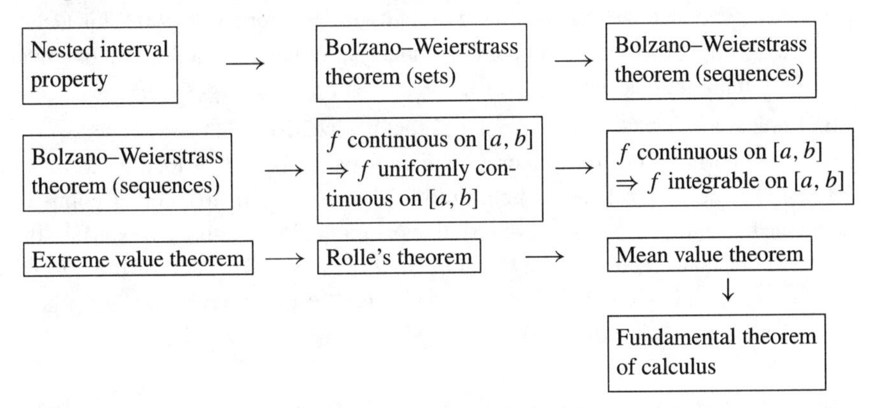

4. State the contrapositive and converse forms of a known theorem.

   Every time we prove a theorem, its contrapositive is a second result we obtain without additional work. For example, after proving the product of convergent sequences must converge, the contrapositive statement that if the sequence $\{x_n\, y_n\}$ diverges, then either $\{x_n\}$ or $\{y_n\}$ must diverge is also true. However, this fact is seldom stated as a separate theorem in analysis texts.

   The exercise of rewriting a theorem in its contrapositive form is another way to reinforce your understanding of the theorem. Some are relatively easy to do, such as the Bolzano–Weierstrass theorem for sets, whose contrapositive is that if a set has no cluster points, then the set is either finite or unbounded. The contrapositive for Rolle's theorem is somewhat more difficult—if there is no value $x \in (a, b)$ for which $f'(x) = 0$, then either (i) $f(a) \neq f(b)$, (ii) f is not differentiable on (a, b), or (iii) f is not continuous at $x = a$ or $x = b$.

   The converse of a given theorem is usually not a true statement. It is good practice to write out converse statements and then look for a counterexample to show why such a statement is not always true. For example, we know that if $\sum_{n=1}^{\infty} a_n$ converges, then $\lim_{n\to\infty} a_n = 0$. The divergent harmonic series

$$\sum_{n=1}^{\infty} \frac{1}{n}$$

   shows that the converse statement is not always true.

5. Look for proof techniques that occur repeatedly.

   One example is the *cancellation principle* (see Problem 3), which is a good technique for proving results such as when a geometric series converges or the funda-

[57] For further practice with this technique, make a flow chart of the convergence tests whose proofs are presented in Essay 8.

mental theorem of calculus. Another example is the practice of "adding zero" to an expression in the form $0 = -b + b$, prior to using the triangle inequality. This is seen in the proof that a product of convergent sequences is convergent, in the proof of the quotient rule for differentiation, and in the proof that a sequence of continuous functions that converges uniformly has a continuous limit function. A third example is to prove that $a = b$ by showing that $|a - b| < \epsilon$ for any $\epsilon > 0$. This technique is used in the proof of the fundamental theorem of calculus and in the interchange of limit theorem that $\lim \int f_n = \int \lim f_n$, when $\{f_n\}$ converges uniformly.

6. Look at a completed proof from different perspectives.

   When the proof of a theorem is completed, it is natural to want to move on to another task. But it can be helpful to look at your proof from other points of view. One possibility is to write a descriptive paragraph that captures the spirit of the proof without concern for all the details. Another is to make a list of the concepts and results used in the proof. This gives an idea of the proof's complexity by the number of other results it requires. Two examples of this are now given.

Example 3. One well-known theorem states that if $f(x)$ is continuous on $[a, b]$, then $f(x)$ is integrable on $[a, b]$. Here is a possible descriptive paragraph about this proof.

In a proof that a continuous function defined on a closed interval is integrable, we use the Riemann integrability criterion. This means that for a given $\epsilon > 0$, we need to find a partition P_ϵ of the interval $[a, b]$ so that the difference between the upper and lower sums $U(f, P_\epsilon) - L(f, P_\epsilon)$ is less than ϵ. To accomplish this, we use the fact that f is uniformly continuous on $[a, b]$ and therefore we have available a δ_ϵ from the definition of uniform continuity. By choosing a partition P_n of $[a, b]$ into n equal parts where $(b - a)/n < \delta_\epsilon$, we are able to show by a chain of inequalities that the Riemann integrability criterion is satisfied.

The second idea is to make a list of the definitions and theorems that are used in the above proof.

Definitions and Basic Facts

1. f is uniformly continuous on $[a, b]$
2. P_ϵ is a partition of the interval $[a, b]$
3. $U(f, P_\epsilon)$ and $L(f, P_\epsilon)$ are upper and lower sums respectively of f for P_ϵ
4. The concepts of $\sup S$ and $\inf S$
5. The Archimedean axiom
6. The fact that $\sum_{i=1}^{n} \Delta x_i = b - a$

Theorems

1. f is continuous on $[a, b]$ implies that f is uniformly continuous on $[a, b]$
2. Extreme value theorem
3. Riemann integrability criterion

Example 4. We next provide a descriptive paragraph about the proof of the Bolzano–Weierstrass theorem for sets, followed by a list of the definitions and theorems used in this proof.

Descriptive paragraph A proof of the Bolzano–Weierstrass theorem that a bounded infinite set must have a cluster point[58] has two parts. In the first part, a nest of closed intervals is constructed by the technique of mathematical induction, in a way that each interval in the sequence is half the length of the previous interval and contains an infinite number of elements from the given set S. The nested interval property then guarantees the existence of a unique value in the intersection of all these intervals. In the second part of the proof, we verify this value is a cluster point of the set by using the definition of a cluster point.

Definitions and Theorems

1. Nested interval property
2. Definition by mathematical induction
3. Definition of a cluster point

Notice that the "list of ingredients" for the Bolzano–Weierstrass theorem is considerably smaller than the one obtained above in Example 3 for the result that a continuous function is integrable.

Summary. Students usually identify the ability to prove theorems as the most difficult part of an analysis course. This skill is not obtained easily or without significant effort. It may be helpful to consider the following eight levels of mathematical understanding.[59] Most students begin an analysis course at level 4. A reasonable goal is to move to level 6 by the end of the course.

1. Being able to do arithmetic.
2. Being able to substitute numbers in formulas.
3. Given formulas, being able to get other formulas.
4. Being able to understand the hypotheses and conclusions of theorems.
5. Being able to understand the proofs of theorems, step by step.
6. Being able to *really* understand the proofs of theorems: that is, seeing why the proof is as it is, and comprehending the inwardness of the theorem and its relation to other theorems.
7. Being able to generalize and extend theorems.
8. Being able to see new relationships and discover and prove entirely new theorems.

[58]Equivalent terms are *limit point* or *point of accumulation*.

[59]These are presented in Underwood Dudley, *Elementary Number Theory*, Second edition, W. H. Freeman and Company, San Francisco, 1978, p. 104.

Essay 10. Sets and Topology

The two important concepts of sequences and sets occur at the beginning of a course in real analysis. Though these are very different concepts, there are enough similarities that students often confuse the two. For example, both sets and sequences can be referred to as bounded, and each can also have a limit point. Further, while sets can be either countable or uncountable, the range set of a sequence is always countable, since the domain consists of the set of natural numbers. A helpful exercise is to list and compare the main results concerning sequences with those for sets.

Point set topology developed in the early part of the 20th century, originally as a tool to understand and generalize some topics in real analysis. As its name suggests, point set topology[60] is concerned with the study of properties of points and sets. These properties are usually introduced at different places (or levels as they are referred to here) to develop a foundation for various topics in analysis. It may help your understanding to collect this material about sets into a single essay.

Level 1. Some basic facts about sets

The basic facts about sets assumed in any mathematics course past calculus include the operations of set union $A \cup B$, set intersection $A \cap B$, and the complement A^c of a given set A. Concepts of the null set $\emptyset$, the universal set X, and a subset of a given set are also part of this basic knowledge. The set difference $A \backslash B$ (or $A - B$) is the set of all elements that are in A but not in B. It is helpful to know that the *power set* of a given set X consists of the collection of all subsets of X. This structure is an example of a Boolean algebra. A *partition* of a given set consists of a collection of non-empty pairwise disjoint subsets whose union is the given set. Examples of a partition are given under Levels 2 and 6. An equivalence relation defined on a set determines a partition of that set into equivalence classes.

Level 2. Some examples of important subsets of the reals $\Re$

The set $\Re$ of real numbers did not receive a careful definition until the last part of the 19th century. This was the final stage in the rigorization of calculus, or the arithmetization of analysis as Weierstrass referred to this program. It began early in the 19th century when Cauchy presented the concept of what we now call the epsilon theory of limits. Most calculus students feel comfortable with the concept of the set of real numbers, also referred to as the real number line, and even the Cartesian product $\Re^2 = \Re \times \Re$, referred to as the Euclidean plane or 2-space.

Several subsets of real numbers are important in their own right. The set of natural numbers **N** (or the positive integers) is fundamental to the definition of a sequence. The prime numbers are a subset of **N** and form the basis for a study of number theory. Two sets that contain **N** are the set **Z** of integers and the set **Q** of rational numbers. A rational number is defined as the quotient of two integers, and can be expressed as a decimal representation

[60] See Joseph W. Dauben, "Set theory and point set topology," in *Companion Encyclopedia of the History and Philosophy of the Mathematical Sciences*, I. Grattan-Guinness, ed., Routledge Inc., New York, 1994, pp. 351–359.

with a repeating pattern. Real numbers that are not rational are called irrational numbers, so the sets of rational and irrational numbers form a partition of the set of real numbers. Another partition of the real numbers consists of the sets of algebraic and transcendental numbers. A number is called *algebraic* if it is the zero of a polynomial with integer coefficients. All rational numbers are algebraic, as are such irrational numbers as $\sqrt{2}$ and $\sqrt[3]{7}$.[61] A real number that is not algebraic is called *transcendental*. Probably the most familiar transcendental numbers are π and e. Note that all transcendental numbers are irrational.

All subsets of real numbers mentioned in the previous paragraph are infinite sets. Towards the end of the 19th century, Georg Cantor developed a theory of infinite sets containing the result that there were two kinds of infinite subsets of $\Re$. The countably infinite sets are those which can be put into a one-to-one correspondence with **N**. Examples of such countable sets are the integers, the rationals, and even the algebraic numbers. Infinite sets, such as the irrationals or transcendental numbers, which cannot be placed into a one-to-one correspondence with the natural numbers, are called uncountably infinite, or uncountable.

Intervals compose another collection of important subsets of $\Re$. Open intervals of the form $(a, b) = \{x \in \Re : a < x < b\}$ serve as neighborhoods of a point and as the basic open set in the usual topology of the reals. For example, an ϵ-neighborhood of the point x is the open interval $(x - \epsilon, x + \epsilon)$. Closed intervals of the form $[a, b] = \{x \in \Re : a \leq x \leq b\}$ serve as the basic closed set in the usual topology of the reals and as the usual domain of functions in a unit on properties of continuous functions. Intervals such as $(a, b]$ or $[a, b)$ are neither open nor closed sets.[62] The unbounded interval (a, ∞) is an open set, while $(-\infty, b]$ is a closed set.

Level 3. Operations defined on a set

A course such as linear algebra or abstract algebra assumes a basic familiarity with the properties of a set as described in Level 1. In addition, these courses present properties of sets with one or two operations defined on members of the set. This study deals with structures such as a vector space, a group, or a field. In a field, we consider properties of operations, such as closure, the associative and commutative laws, and the existence of an identity and inverse elements. The most important set for the study of analysis is the set of real numbers, which is the prototype of a complete ordered field. The adjectives of ordered and complete are described next.

Level 4. An order relation defined on a set

One of the first concepts presented in an analysis course is that of a bounded set of real numbers, with the related terms of infimum (or greatest lower bound) and supremum (or least upper bound) for such a set. These are needed for a statement of the *least upper bound axiom*, also called the *completeness axiom* or *supremum principle*. This axiom states that every non-empty set with an upper bound must have a least upper bound, and is used to

[61] $\sqrt{2}$ is a zero of $x^2 - 2 = 0$, while $\sqrt[3]{7}$ is a zero of $x^3 - 7 = 0$.

[62] Open and closed sets are defined in Level 6.

prove several important results about sets and sequences, such as the Bolzano–Weierstrass theorems.[63]

In order to define the concept of a bounded set, we need an order relation, so that a statement such as "$x < b$ for all $x \in S$" will have meaning. The less than ($<$) order relation satisfies the transitive property and the trichotomy law, which makes $(\mathfrak{R}, <)$ a simply ordered set. The order relation ($<$) and the operations $(+, \cdot)$ can be combined to prove identities such as: if $x < y$, then $a + x < a + y$ for all a, $ax < ay$ for $a > 0$, and $ax > ay$ for $a < 0$.

Level 5. Consequences of the completeness axiom

These consequences occur in the discussion of sets, sequences, and series. For sets, we first prove the Archimedean axiom which states: if $x > 0$, there is some $n \in \mathbf{N}$ so that $0 < 1/n < x$. This result enables us to prove the density theorem which states that between any two real numbers, there exist infinitely many rational and irrational numbers. Actually, the proof shows there exists a rational and an irrational between two given real numbers, but if there is one, we can prove there must be a countably infinite number of them.[64]

Another result is the nested interval property, which asserts that any sequence $\{I_n\}$ of closed intervals where $I_{n+1} \subset I_n$ and $\lim_{n\to\infty} \ell(I_n) = 0$ must have exactly one real number that is in every interval I_n. This property provides one possible approach for proving the important Bolzano–Weierstrass theorem for sets. Also needed is the definition of a cluster point for a set. We say that x is a *cluster point* for the set S if every neighborhood of x contains a point of S different from x itself. Another way to state this is that every neighborhood U of x satisfies

$$U \cap (S - \{x\}) \neq \emptyset.$$

The designation of a limit point for a set or an accumulation point may be used in place of cluster point. The derived set S' consists of all cluster points of the set S.

We next consider some consequences of the completeness axiom for sequences. After introducing the concept of a limit point for a sequence (i.e., x is a *limit point* for the sequence $\{x_n\}$ if $\{x_n\}$ has a subsequence that converges to x), the Bolzano–Weierstrass theorem for sets can be used to prove the Bolzano–Weierstrass theorem for sequences. This important result that a bounded sequence must have a convergent subsequence is used in the proofs of many continuity theorems.

The completeness axiom is also used to prove that a bounded and monotone sequence converges. This result, in turn, is fundamental in the proofs of all the convergence tests for infinite series of positive terms (see Essay 8). It is also used to prove that the sequence

$$\{e_n\} \quad \text{where } e_n = \left(1 + \frac{1}{n}\right)^n$$

converges. This sequence is well-known since it converges to the number e.

---

[63]The Bolzano–Weierstrass theorem for sets states that a bounded, infinite set must have a cluster point. The Bolzano–Weierstrass theorem for sequences states that a bounded sequence must have a convergent subsequence, and therefore a limit point.

[64]It is also true that there are infinitely many irrational numbers between two given rational numbers.

Level 6. Open and closed sets, and compactness

To this point, we have considered the concepts of a cluster point for a set, a limit point for a sequence, and a set that is bounded and/or infinite. The two Bolzano–Weierstrass theorems are the main results obtained from these ideas. At this new level, we add the definitions for interior, exterior, and boundary points of a set, along with the definitions for a set to be open, closed, or compact. These additional topological ideas are used to express properties needed for proofs of the continuity theorems. We begin with three definitions concerning points.

> x is an *interior point* of the set S if there is a neighborhood U of x so that $U \subset S$ (or equivalently, if there is some $\epsilon > 0$ so that if $x - \epsilon < y < x + \epsilon$, then $y \in S$).
>
> x is an *exterior point* of the set S if there is a neighborhood U of x so that $U \subset \Re - S$ (or if $U \cap S = \emptyset$).
>
> x is a *boundary point* of the set S if it is neither an interior point nor an exterior point of S (if U is a neighborhood of a boundary point x, then $U \cap S \neq \emptyset$ and $U \cap (\Re - S) \neq \emptyset$.)

Based on these definitions concerning points, we define three new sets as follows: int S is the set of all interior points of S, ext S is the set of all exterior points of S, and bd S is the set of all boundary points of S. Notice that these three sets form a partition of the reals $\Re$, if we omit any sets that are empty. As an example, if $S = (a, b]$, then

$$\text{int } S = (a, b), \quad \text{ext } S = (-\infty, a) \cup (b, \infty), \quad \text{and} \quad \text{bd } S = \{a\} \cup \{b\}.$$

Observe also that the derived set is $S' = [a, b]$.

A final step is to define what is meant by an open set, a closed set, and a compact set. Each of these three types of sets can be characterized by two different criteria. Depending on which book you read, one criterion is chosen as the definition and the other as a theorem. The first entry below for each is the one most frequently chosen as the definition.

A is an *open set*	if and only if every point in A is an interior point of A if and only if $\Re - A$ is a closed set.
B is a *closed set*	if and only if $\Re - B$ is an open set if and only if B contains all its cluster points.
C is a *compact set*	if and only if every open cover of C contains a finite subcover if and only if C is closed and bounded.

In the example of $S = (a, b]$ given above, S is not an open set because $S \subseteq \text{int } S$ does not hold. S is not a closed set because $S' \subseteq S$ does not hold.[65]

The definition of compactness requires some discussion of what is meant by an open cover and a finite subcover of a set. The collection $\mathcal{G}$ of open sets is an *open cover* for the set S if $S \subset \cup_{G \in \mathcal{G}} G$. By this we mean that S is contained in the union of all the open sets in the collection $\mathcal{G}$. This collection has a *finite subcover* if there are $G_1, G_2, \ldots, G_n$ in $\mathcal{G}$ so that $S \subset \cup_{i=1}^{n} G_i$.

[65] See Problem 8 for practice with these new concepts.

The first important use of these concepts is in proofs of continuity theorems (such as the extreme value theorem) where we assume the domain D of the continuous function f is a compact set. Because D is a bounded set, we can use the Bolzano–Weierstrass theorem to obtain a cluster point of D. Because D is a closed set, we know that any such cluster points must be in D. These are essential facts for the proofs of the continuity theorems.

Level 7. More involved topological concepts

By introducing more involved concepts, such as those of connected sets, perfect sets, and separable sets, we develop more properties that are part of topology, but which are not usually needed for a first course of real analysis.

Selected Readings

Introduction

The purpose of this section is to provide a sample choice of readings to supplement and enrich material covered in a traditional analysis course. The readings are chosen to present a variety of topics from several time periods.

Selection 1. John Bernoulli and the Marquis de l'Hôpital

A standard story for calculus students is how l'Hôpital who knew very little mathematics "stole" the knowledge of his rule from John Bernoulli. The beginning of this selection describes the first meeting of these two men in 1691 from the viewpoint of John Bernoulli. At this time, John Bernoulli was 24 years of age, while l'Hôpital was 30. The second part is in l'Hôpital's words from the Preface of his text on derivatives which was published in 1696. This was the first calculus text to be written, and the author did a creditable job. In it, l'Hôpital gives full credit for the discoveries of the Bernoullis. So perhaps the standard story needs some modification. The name l'Hôpital is written in the more usual form of L'Hospital throughout the rest of this text.

Selection 2. The Euler–Stirling Correspondence

Very little of Euler's writings has been translated into English, so these letters provide an unusual opportunity to gain insights into the life and work of Leonhard Euler. Two of his letters are included in this selection along with a reply from James Stirling. They are written near the beginning of Euler's career at the Academy in St. Petersburg, and contain glimpses into some of his early mathematical interests and results. The letters also provide a better picture of the personal qualities of Euler than can be found in his mathematical writings. Places where some of the more technical material is omitted are indicated by [...]. See the original text for an extensive set of footnote comments which are omitted from this selection. See also Problem 28.

Selection 3. Developing Rigor in Calculus

Judith Grabiner wrote the book *The Origins of Cauchy's Rigorous Calculus* (see **[18]**) about the introduction of rigor into calculus and analysis, emphasizing the efforts of Cauchy. She has also written articles and given talks on this topic. This selection is one of her articles that was published by the *American Mathematical Monthly* in 1974.

Selection 4. The Contribution of Lebesgue

The *Biographical Dictionary of Mathematicians* (see **[16]**) is an excellent source for information about the life and work of virtually all major mathematicians. This is especially helpful since there are full-length biographies for relatively few mathematicians. In this selection, Thomas Hawkins describes the life and influence of Henri Lebesgue, the French mathematician in the early twentieth century who developed the important tools of the Lebesgue measure and integral. His main discovery was contained in his doctoral dissertation—one of the rare instances when a doctoral dissertation contains a significant mathematical result.

Selection 5. The Bernoulli Numbers and Some Wonderful Discoveries of Euler

Students in calculus learn how to find several terms of the Maclaurin series for $\tan x$ by use of long division, but do not know how to find an expression for the general term. They may also be aware that the p-series

$$\sum_{n=1}^{\infty} \frac{1}{n^2}$$

converges to

$$\frac{\pi^2}{6},$$

and think of this as an isolated result. This selection first develops a recurrence relation for the Bernoulli numbers, and then provides an outline of steps to show how these Bernoulli numbers occur in the general term of the power series for $\tan x$ as well as in the value for

$$\sum_{n=1}^{\infty} \frac{1}{n^{2k}}.$$

You will find it enjoyable and satisfying to fill in the details.

Selection 1. John Bernoulli and the Marquis de l'Hôpital

The History of Mathematics: A Reader
John Fauvel and Jeremy Gray, editors
MacMillan Press, 1987
pages 441–442
Reprinted by permission of Palgrave MacMillan

13.B4 O. Spiess on Bernoulli's First Meeting with l'Hôpital

Immediately on arriving in Paris in late Autumn 1691, (Johann) Bernoulli visited Father Malebranche who once a week played host to the best known scholars of the city. Rather as an admission card, so to speak, he showed the famous philosophers, who were also good mathematicians, his construction of the catenary—on a single piece of paper—that had just been published a little earlier in the June volume of the *Acta Eruditorum* and which was his first important achievement. Because of this he was invited by Malebranche to take part regularly at his meetings. So the 24-year-old student appeared in the illustrious circle on the next occasion and was immediately introduced to the Marquis de l'Hôpital, to whom Malebranche had showed his paper a few days before.

'From the conversation' [Bernoulli wrote to his friend Montmort in 1718] 'which I had with M. le Marquis I knew right away that he was a good geometer for what was already known, but that he knew nothing at all of the differential calculus, of which he scarcely knew the name, and still less had he heard talk of the integral calculus in the *Acta* of Leipzig having not yet reached him because of the war.' The Marquis, who found it difficult to see the conqueror of the catenary in the young man, examined him backwards and forwards 'but he saw soon enough that I was neither an adventurer nor the pretender that he believed I wanted to play at being. The conversation finally fell to the developed curve [evolute] or osculating circle, for the study of which he prided himself on an entirely particular rule drawn from M. Fermat's method of max. and min. To test him, I proposed an example of an algebraic curve (for this supposedly general rule only worked for algebraic curves and only gave the radius at the maximum).'

Mr l'Hôpital took paper and ink and began to calculate. After he had used up nearly an hour in scribbling over several pieces of paper he finally found the correct value of the radius at the maximum of the curve.'

Bernoulli then said to him that there was a formula that would find the radius of curvature of any curve at any point in a few minutes, and put forward a curve for which he could find the sought-for value at once 'which struck him so much with surprise that from that moment he became charmed with the new analysis of the infinitely small and excited with the desire to learn it from me.' Bernoulli visited the Marquis the very next day, who asked him to visit four times a week 'to explain to him on each occasion and then to deliver a lesson based on the paper which I had written at home the evening before'. And, what is of particular importance now 'one of my friends from Basel who was lodging with me had the kindness to copy each of the papers I was to take to M. le Marquis, so I have preserved them all'. [...]

The lessons in Paris went on from the end of 1691 to the end of July 1692, so for over half a year; then l'Hôpital took his young instructor to his estate in Oucques where Bernoulli presented his lectures in daily contact with the Marquis and his spirited wife. 'I didn't hesitate' he wrote in the letter to Montmort 'to give to M. l'Hopital new memoirs always written in my own hand whenever I found appropriate material, and he furnished me himself with the occasion for all sorts of questions.'

13.B5 Preface to l'Hôpital's *Analyse des Infiniment Petits*

The Defect of this Method was supplied by that of Mr Leibnitz's (footnote by l'Hôpital *Acta Erudit. Lips. Ann. 1684, p. 467*) [footnote by Stone: rather the great Sir Isaac Newton—see Commercium Epistolicum]. He began where Dr Barrow and others left off. His Calculus has carried him into Countries hitherto unknown; and he has made Discoveries by it astonishing the greatest Mathematicians of Europe. The Messieurs Bernoulli were the first who perceived the Beauty of the Method; and have carried it to such a length, as by its means to surmount Difficulties that were before thought insuperable.

I intended to have added another Section to shew the surprising use of this calculus in Physicks, and to what degree of Exactness it may bring the same; as likewise the use thereof in Mechanicks; But Sickness has prevented me herein. However, I hope to effect it hereafter, and present it the Publick with interest. And indeed the whole of the present Treatise is only the First Part of the Calculus of Mr Leibnitz, or the Direct Method, wherein we descend from Whole Magnitudes to their infinitely small Parts of what kind soever comparing them with each other, which is called the Calculus Differentialis: But the other Part, called the Calculus Integralis, [or Inverse Method of Fluxions] consists on ascending from these infinitely small Parts to the Magnitudes, or Wholes, whereof they are the Parts. This Inverse Method I also designed to publish but Mr. Leibnitz's having wrote to me, that he was at work upon this subject, in order for a Treatise de Scientia Infiniti, I was unwilling to deprive the Publick of so fine a Piece, which needs contain whatever is curious in the Inverse Method of Tangents, Rectifications of Curves, Quadratures, Investigation of Superficies of Solids, and their Solidities, Centres of Gravity, etc. Neither would I ever have published the present Treatise, had he not intreated me to it by Letter; as likewise because I believed it might prove a necessary Introduction to whatever should hereafter be discovered on the subject.

I must own myself very much obliged to the labours of Messieurs Bernoulli, but particularly to those of the present Professor at Groeningen, as having made free with their Discoveries as well as those of Mr Leibnitz: So that whatever they please to claim as their own I frankly return them.

I must here in justice own (as Mr Leibnitz himself has done, in *Journal des Scavans* for August 1694) that the learned Sir Isaac Newton likewise discovered something like the Calculus Differentialis, as appears by his excellent *Principia*, published first in the Year 1687 which almost wholly depends on the Use of the said Calculus. But the Method of Mr. Leibnitz's is much more easy and expeditious, on account of the Notation he uses, not to mention the wonderful assistance it affords on many occasions.

Selection 2. The Euler–Stirling Correspondence

James Stirling
by Ian Tweddle
Scottish Academic Press, Edinburgh, 1987
pages 140–154

Parts of these letters and their translations by Ian Tweddle are reprinted by permission of the author. [. . .] indicates where there are omissions.

Introduction by Ian Tweddle

In a letter of 16 November 1734, Euler requested a Danish naval officer, Friedrick Weggersloff to make contact with the Royal Society in London with a view to establishing a regular channel of communication through which Euler could exchange ideas with the mathematicians of the Royal Society. Weggersloff's task was frustrated by an illness and presumably also by the brevity of his visits to London but eventually he was able to write to Euler on 16 March 1736 that Stirling would be honoured to correspond with Euler and that the correspondence could be in Latin or French. Euler was obviously delighted to make contact with Stirling in this way but unfortunately his expectations of the association were not fulfilled. With Stirling increasingly involved in his duties at Leadhills, the contact was broken before it became properly established. Euler had to wait almost two years before he received a reply to his first letter while his second letter appears to have gone unanswered. [. . .]

Euler's First Letter to Stirling Dated 8 June 1736 at Petrograd

For a long time I have burned with a desire to set up an exchange of letters with an eminent mathematician of your country, not only that I might benefit from your exceptional discoveries but especially because I would give great satisfaction to our Academy as a result. In this matter I owe the greatest debt of gratitude to the most learned Wegersloff, because he has put me in contact with you above all: for the more I have learned from your excellent articles, which I have seen here and there in your Transactions, concerning the nature of series, a study in which I have indeed expended much effort, the more I have wished to become acquainted with you in order that I could receive more from you yourself and also submit my own deliberations to your judgement. But before I wrote to you, I searched all over with great eagerness for your excellent book on the method of differences, a review of which I had seen a short time before in the *Actae Lipslienses*, until I achieved my desire. Now that I have read through it diligently, I am truly astonished at the great abundance of excellent methods contained in such a small volume, by means of which you show how to sum slowly converging series with ease and how to interpolate progressions which are very difficult to deal with. But especially pleasing to me was prop. XIV of Part I in which you give a method by which series, whose law of progressions is not even established, may be summed with great ease using only the relation of the last terms; certainly this method extends very widely and is of the greatest use. In fact the proof of this proposition, which you seem to have deliberately withheld, caused me enormous difficulty, until at last

I succeeded with very great pleasure in deriving it from the preceding results, which is the reason why I have not yet been able to examine in detail all the subsequent propositions. Concerning the summation of very slowly converging series, in the past year I have lectured to our Academy on a special method by means of which I have given the sums of very many series sufficiently accurately and with very little effort. Namely let

$$S = A^1 + B^2 + C^3 \ldots X^x,$$

where the superscripts denote the position of each term in the series. Now the sum of these terms from the first one A up to any term X is

$$\begin{aligned} &= \int X\,dx + \frac{X}{1\cdot 2} + \frac{dX}{1\cdot 2\cdot 3\cdot 2\,dx} - \frac{d^3X}{1\cdot 2\cdot 3\cdot 4\cdot 5\cdot 6\,dx^3} \\ &+ \frac{d^5X}{1\cdot 2\cdot 3\cdot 4\cdot 5\cdot 6\cdot 7\cdot 6\,dx^5} - \frac{3d^7X}{1\cdot 2\cdots 9\cdot 10\,dx^7} + \frac{5d^9X}{1\cdot 2\cdots 11\cdot 6\,dx^9} \\ &- \frac{691d^{11}X}{1\cdot 2\cdots 13\cdot 210\,dx^{11}} + \frac{35d^{13}X}{1\cdot 2\cdots 15\cdot 2\,dx^{13}} - \frac{3617\,d^{15}X}{1\cdot 2\cdots 17\cdot 30\,dx^{15}} \\ &+ \frac{2423279\,d^{17}X}{1\cdot 2\cdots 19\cdot 1890\,dx^{17}} - \text{etc.} \end{aligned} \tag{1}$$

In this expression when X is assumed to be given in terms of x both $\int X\,dx$ and all the remaining terms will be assignable; moreover in writing these terms I am compelled by necessity to use the Leibniz notation since it would have been inconvenient to write 17 points above X in place of $d^{17}X$. Then since the integration of $\int X\,dx$ allows the addition of a constant quantity, it must be so determined that $S = 0$ if one puts $x = 0$. Therefore by means of this formula all series may be summed as far as desired and that can be done exactly if the higher order fluxions of X eventually vanish as happens if X is a positive power of x or an aggregate of powers of this type.

Thus if the sum of the progression

$$1 + 2^{12} + 3^{12} + 4^{12} + 5^{12} + \cdots + x^{12}$$

is required, from $X = x^{12}$ we will have

$$\int X\,dx = x^{13}/13; \quad dX/dx = 12x^{11};$$

$$d^3X/dx^3 = 12\cdot 11\cdot 10x^9;$$

$$d^5X/dx^5 = 12\cdot 11\cdot 10\cdot 9\cdot 8x^7$$

and finally

$$d^{11}X/dx^{11} = 12\cdot 11\cdot 10\cdots 2x,$$

all the subsequent terms vanishing. Therefore when these terms have been put together, the sum will come out as

$$S = \frac{x^{13}}{13} + \frac{x^{12}}{2} + x^{11} - \frac{11x^9}{6} + \frac{22x^7}{7} - \frac{33x^5}{10} + \frac{5x^3}{3} - \frac{691x}{2730}.$$

But when the given series terminates at no point, then the approximate sum may at least be found. Thus if the series under consideration is

$$1 + \frac{1}{2} + \frac{1}{3} + \frac{1}{4} + \cdots + \frac{1}{x},$$

one finds the sum of these terms

$$S = C + lx + \frac{1}{2x} - \frac{1}{12x^2} + \frac{1}{120x^4} - \frac{1}{252x^6} + \frac{1}{240x^8} - \frac{1}{132x^{10}} + \frac{691}{32760x^{12}} - \text{etc.}$$

However in this case it is difficult to define the constant C by the prescribed method. For this purpose I put $x = 10$, and add ten terms directly, the sum of which will be

$$S = 2.9289682539682539;$$

whence we will have

$$C = 0.5772156649015329.$$

And having found this constant one may readily assign the sum of any number of terms of that series. Thus I have found the sum of a thousand terms

$$= 7.4854708605503449$$

and the sum of a thousand thousand terms

$$= 14.3927267228657236. [\ldots] \tag{2}$$

In a similar way we obtain

$$S = 1 + \frac{1}{4} + \frac{1}{9} + \frac{1}{16} + \cdots + \frac{1}{x^2}$$

$$= C - \frac{1}{x} + \frac{1}{2x^2} - \frac{1}{6x^3} + \frac{1}{30x^5} - \frac{1}{42x^7} + \frac{1}{30x^9} - \frac{5}{66x^{11}} + \text{etc.} \tag{3}$$

In fact the constant C, which may be conveniently found by adding up some terms directly, is equal to the sum of the series continued to infinity, that is 1.64493406684822643647 which agrees especially well with the value which you have given. Moreover I have found by another method that the sums of series of reciprocal powers in which the exponents are even, all depend on the quadrature of the circle, for if the ratio of the diameter to the circumference is taken as 1 to p then the results are as follows

$$1 + \frac{1}{2^2} + \frac{1}{3^2} + \frac{1}{4^2} + \frac{1}{5^2} + \text{etc.} = \frac{1}{6}p^2 \tag{4}$$

$$1 + \frac{1}{2^4} + \frac{1}{3^4} + \frac{1}{4^4} + \frac{1}{5^4} + \text{etc.} = \frac{1}{90}p^4$$

$$1 + \frac{1}{2^6} + \frac{1}{3^6} + \frac{1}{4^6} + \frac{1}{5^6} + \text{etc.} = \frac{1}{945}p^6$$

$$1 + \frac{1}{2^8} + \frac{1}{3^8} + \frac{1}{4^8} + \frac{1}{5^8} + \text{etc.} = \frac{1}{9450}p^8$$

This property seems to me to be worthy of further consideration because it cannot be established by any of the accepted methods.

Stirling's Reply Dated 16 April 1738 at Edinburgh

So much time has elapsed since you were so kind to write to me, that I would have scarcely dared to reply now if I did not rely upon your courtesy. During these last two years I have been involved in a great many business matters which have required me to go frequently to Scotland and then return to London. And it was on account of these affairs that first of all your letter came late into my hands and then that, even to this very day, there is scarcely time available for reading through your letter with the attention which it deserves. For after deliberations have been interrupted, not to say neglected, for a long time, patience is required before the mind can be brought to think about the same things once again. Therefore I seize this first opportunity to express my respects to you and at the same time to thank you belatedly for your letter which was filled with exceptional discoveries.

Most pleasing to me was your theorem for summing series by means of the area of a curve and derivatives or fluxions of terms for it is general and well-suited for application. I immediately perceived of extending it to very many types of series, and what is extraordinary, it approximates very rapidly in most cases. Perhaps you have not noticed that my theorem for summing logarithms is nothing more than a special case of your general theorem. But this discovery was all the more pleasing to me because I had also thought about the same matter some time ago; but I did not proceed beyond the first term, and using only it I approximated as I wished to the values of series with sufficient ease just by repeating the calculation as in the solution of affected equations; I have given an example of this in our Philosophical Transactions. [...]

But most pleasing of all for me was your method for summing certain series by means of powers of the circumference of the circle. I acknowledge this to be quite ingenious and entirely new and I do not see that it has anything in common with the accepted methods, so that I readily believe that you have drawn it from a new source.

Your series are contained in the general form

$$1 + \frac{1}{2^n} + \frac{1}{3^n} + \frac{1}{4^n} + \frac{1}{5^n} + \frac{1}{6^n} + \text{etc.}$$

This is reduced with no effort to the following form,

$$1 + \frac{1}{3^n} + \frac{1}{5^n} + \frac{1}{7^n} + \frac{1}{9^n} + \frac{1}{11^n} + \text{etc.}$$

And you show how to sum this by means of the nth power of the circumference whenever n is even. Moreover if the signs of alternate terms are changed so that the series becomes

$$1 - \frac{1}{3^n} + \frac{1}{5^n} - \frac{1}{7^n} + \frac{1}{9^n} - \frac{1}{11^n} + \text{etc.}$$

I say that this can always be summed by the nth power of the circumference whenever n is odd; in particular if

$$n=1, \qquad \frac{1}{4}p = 1 - \frac{1}{3} + \frac{1}{5} - \frac{1}{7} + \frac{1}{9} - \frac{1}{11} + \text{etc. as is generally well-known}$$

$$n=3, \qquad \frac{1}{32}p^3 = 1 - \frac{1}{3^3} + \frac{1}{5^3} - \frac{1}{7^3} + \frac{1}{9^3} - \frac{1}{11^3} + \text{etc.}$$

$$n=5, \qquad \frac{5}{1536}p^5 = 1 - \frac{1}{3^5} + \frac{1}{5^5} - \frac{1}{7^5} + \frac{1}{9^5} - \frac{1}{11^5} + \text{etc.}$$

I do not doubt that you have already noted the same result; otherwise it is easily derived from your basic principle, which I will be pleased to see if it seems appropriate to you.

At this point you should be advised that in due course Mr MacLaurin, Professor of Mathematics at Edinburgh, will be publishing a book on fluxions. He has communicated to me some of its pages which have already been printed. In these he has two theorems for summing series by means of derivatives of the terms, one of which is the self-same result that you sent me; I have informed him of this. Although he had willingly promised that he would acknowledge this in his preface, I nevertheless submit to your judgement whether you wish to publish your letter in our Philosophical Transactions. If you wish to illustrate or prove certain things, and write back to me quickly, I shall ensure that it sees the light of day before his book comes out. For if you should have a desire to be elected one of the Fellows of our Royal Society at this time, there is no doubt that this will be agreeable to the others when they see your most splendid discoveries.

Euler's Second Letter to Stirling Dated 27 July 1738 at Petrograd

The greater the longing with which I have awaited a letter from you, the greater the joy which your most kind reply has brought me and I am all the more delighted with it because I see not only that my letter was not unpleasing to you but also that you yourself invite the continuation of this exchange which has been initiated. I am therefore most grateful to you because you have been willing to receive my meagre thoughts so kindly and to communicate your judgement of them to me. Moreover I ascribe it to your great kindness that my letter is judged by you to be worthy of inclusion in your Transactions, and it seemed appropriate for this purpose to add several extensions and clarifications, which you may add or omit as you see fit. But in this matter I have very little desire for anything to be detracted from the fame of the celebrated Mr MacLaurin since he probably came upon the same theorem for summing series before me, and consequently deserves to be named as its first discoverer. For I found that theorem about four years ago, at which time I also described its proof and application in greater detail to our Academy; my dissertation on this along with the one I have composed on the summation of series by means of powers of the circumference of the circle will shortly see public light in our *Commentarii* which come out each year. Moreover in our *Commentarii* which have now been published there are some other methods of mine for summing series, certain of which have great similarity to your methods which are presented in your distinguished work, but because I had not yet seen your method of differences at that time, I was also unable to make the due reference to it. A good many years ago I also sent to your most illustrious president Mr Sloane a certain manuscript in which I gave the general solution of this equation $\dot{y} = y^2\dot{x} + ax^m\dot{x}$ which had previously been thought about a great deal but had been solved for only very few cases of the exponent m. Therefore if this dissertation were still to hand, it might be presented to your Society as an example of my work when it has judged me worthy of membership, an honour for which I should be indebted to you alone. But I am afraid that it may not be advantageous for the renowned Society to elect me as a Fellow since I am so severely restrained at our Academy that my thoughts of whatever nature have to be presented here first of all.

But to return to the theorem by which the sum of any series can be found when its general term has been stated, it is clear that the given form would become more useful as more of its terms are obtained, but it seems to be extremely difficult to continue it arbitrarily. Indeed I have not extended it to more than twelve terms, the last of which I have found not so very long ago; now this expression is as follows.

If the first term of any series is A, the second B, the third C, etc. and the one whose index is x is $= X$: the sum of this progression will be

$$\begin{aligned} A + B + C + \text{ etc. } \cdots + X = \int X\,dx + \frac{X}{1\cdot 2} + \frac{dX}{1\cdot 2\cdot 3\cdot 2dx} \\ - \frac{d^3X}{1\cdot 2\cdot 3\cdot 4\cdot 5\cdot 6\,dx^3} + \frac{d^5X}{1\cdot 2\cdot 3\cdot 4\cdot 5\cdot 6\cdot 7\cdot 6\,dx^5} \\ - \frac{3d^7X}{1\cdot 2\cdot 3\cdot\cdots 9\cdot 10\,dx^7} + \frac{5\,d^9X}{1\cdot 2\cdot 3\cdots 11\cdot 6\,dx^9} \\ - \frac{691\,d^{11}X}{1\cdot 2\cdot 3\cdots 13\cdot 210\,dx^{11}} + \frac{35\,d^{13}X}{1\cdot 2\cdot 3\cdots 15\cdot 2\,dx^{13}} \\ - \frac{3617d^{15}X}{1\cdot 2\cdot 3\cdots 17\cdot 30\,dx^{15}} + \frac{43867d^{17}X}{1\cdot 2\cdot 3\cdots 19\cdot 42\,dx^{17}} \\ - \frac{1222277d^{19}X}{1\cdot 2\cdot 3\cdots 21\cdot 110dx^{19}} \text{ etc.} \end{aligned} \tag{5}$$

where the fluxion dx is set constant. Moreover with slight changes this expression can be adapted for finding the sum of a series from the term X up to infinity. In addition to the marked facility which it provides for finding sums approximately, this form is exceptionally useful in the investigation of the true sums of algebraic series, whose sums can be set out completely; thus if the sum of this progression of powers is sought

$$1 + 2^{12} + 3^{12} + 4^{12} + 5^{12} + \cdots + x^{12},$$

we will have

$$X = x^{12}, \quad \int X\,dx = \frac{1}{13}x^{13}, \quad \frac{dX}{dx} = 12x^{11}, \quad \frac{d^3X}{dx^3} = 10\cdot 11\cdot 12\cdot x^9,$$

and so on until $d^{13}X/dx^{13}$ along with the following terms $= 0$.

Hence the required sum will come out

$$= \frac{x^{13}}{13} + \frac{x^{12}}{2} + x^{11} - \frac{11x^9}{6} + \frac{22x^7}{7} - \frac{33x^5}{10} + \frac{5x^3}{3} - \frac{691x}{2730},$$

and I do not know if this sum can be found so easily by any other method. But using this procedure one may find just as conveniently the sum of this progression

$$1 + 2^{21} + 3^{21} + 4^{21} + \cdots + x^{21},$$

while the work involved with other methods seems insurmountable. [...]

Next the material which pertains to the summation of series of the type contained in this form:

$$1 + \frac{1}{2^n} + \frac{1}{3^n} + \frac{1}{4^n} + \frac{1}{5^n} + \text{etc.}$$

where n is an even number—I have tackled these by two methods, one of which, as you correctly inferred, I have deduced from the series

$$1 + \frac{1}{3^n} + \frac{1}{5^n} + \frac{1}{7^n} + \text{etc.},$$

while the other has provided me with the sum of the former series directly. By the first method in particular, I have also determined the sums of series of this type

$$1 - \frac{1}{3^n} + \frac{1}{5^n} - \frac{1}{7^n} + \frac{1}{9^n} - \text{etc.},$$

where n is an odd number, and I have found these to be just as you point out. Moreover the sums for both even and odd exponents n are the following.

$$\begin{aligned}
\frac{p}{4} &= 1 - \frac{1}{3} + \frac{1}{5} - \frac{1}{7} + \frac{1}{9} - \text{etc.} \\
\frac{p^2}{8} &= 1 + \frac{1}{3^2} + \frac{1}{5^2} + \frac{1}{7^2} + \frac{1}{9^2} + \text{etc.} \\
\frac{p^3}{32} &= 1 - \frac{1}{3^3} + \frac{1}{5^3} - \frac{1}{7^3} + \frac{1}{9^3} - \text{etc.} \\
\frac{p^4}{96} &= 1 + \frac{1}{3^4} + \frac{1}{5^4} + \frac{1}{7^4} + \frac{1}{9^4} + \text{etc.} \\
\frac{5p^5}{1536} &= 1 - \frac{1}{3^5} + \frac{1}{5^5} - \frac{1}{7^5} + \frac{1}{9^5} - \text{etc.} \\
\frac{p^6}{960} &= 1 + \frac{1}{3^6} + \frac{1}{5^6} + \frac{1}{7^6} + \frac{1}{9^6} + \text{etc.} \\
\frac{61p^7}{184320} &= 1 - \frac{1}{3^7} + \frac{1}{5^7} - \frac{1}{7^7} + \frac{1}{9^7} - \text{etc.} \\
\frac{17p^8}{161280} &= 1 + \frac{1}{3^8} + \frac{1}{5^8} + \frac{1}{7^8} + \frac{1}{9^8} + \text{etc.}
\end{aligned} \tag{6}$$

all those series being contained in this one general form:

$$1+\left(-\frac{1}{3}\right)^n+\left(+\frac{1}{5}\right)^n+\left(-\frac{1}{7}\right)^n+\left(+\frac{1}{9}\right)^n+\text{etc.}$$

where n is an integer. For if n is an even number, then all the terms will have the sign $+$; but if on the other hand n is odd, then the signs follow one another alternately. [...]

Moreover in those sums there is a striking relationship between their numerical coefficients and the terms of the above progression which I gave first of all for summing any series, namely

$$\int X\,dx+\frac{X}{1\cdot 2}+\frac{dX}{1\cdot 2\cdot 3\cdot 2dx}-\text{etc.},$$

which deserves to be noted. It seemed useful to reproduce the individual sums in the corresponding way so that this relationship may be seen more clearly.

$$\frac{2^1\cdot 1}{1\cdot 2\cdot 3\cdot 2}p^2=1+\frac{1}{2^2}+\frac{1}{3^2}+\frac{1}{4^2}+\frac{1}{5^2}+\text{etc.} \tag{7}$$

$$\frac{2^3\cdot 1}{1\cdot 2\cdot 3\cdot 4\cdot 5\cdot 6}p^4=1+\frac{1}{2^4}+\frac{1}{3^4}+\frac{1}{4^4}+\frac{1}{5^4}+\text{etc.}$$

$$\frac{2^5\cdot 1}{1\cdot 2\cdot 3\cdot 4\cdot 5\cdot 6\cdot 7\cdot 6}p^6=1+\frac{1}{2^6}+\frac{1}{3^6}+\frac{1}{4^6}+\frac{1}{5^6}+\text{etc.}$$

$$\frac{2^7\cdot 3}{1\cdot 2\cdot 3\cdots 9\cdot 10}p^8=1+\frac{1}{2^8}+\frac{1}{3^8}+\frac{1}{4^8}+\frac{1}{5^8}+\text{etc.}$$

$$\frac{2^9\cdot 5}{1\cdot 2\cdot 3\cdots 11\cdot 6}p^{10}=1+\frac{1}{2^{10}}+\frac{1}{3^{10}}+\frac{1}{4^{10}}+\frac{1}{5^{10}}+\text{etc.}$$

$$\frac{2^{11}\cdot 691}{1\cdot 2\cdot 3\cdots 13\cdot 210}p^{12}=1+\frac{1}{2^{12}}+\frac{1}{3^{12}}+\frac{1}{4^{12}}+\frac{1}{5^{12}}+\text{etc.}$$

$$\frac{2^{13}\cdot 35}{1\cdot 2\cdot 3\cdots 15\cdot 2}p^{14}=1+\frac{1}{2^{14}}+\frac{1}{3^{14}}+\frac{1}{4^{14}}+\frac{1}{5^{14}}+\text{etc.}$$

$$\frac{2^{15}\cdot 3617}{1\cdot 2\cdot 3\cdots 17\cdot 30}p^{16}=1+\frac{1}{2^{16}}+\frac{1}{3^{16}}+\frac{1}{4^{16}}+\frac{1}{5^{16}}+\text{etc.}$$

$$\frac{2^{17}\cdot 43867}{1\cdot 2\cdot 3\cdots 19\cdot 42}p^{18}=1+\frac{1}{2^{18}}+\frac{1}{3^{18}}+\frac{1}{4^{18}}+\frac{1}{5^{18}}+\text{etc.}$$

$$\frac{2^{19}\cdot 1222277}{1\cdot 2\cdot 3\cdots 21\cdot 110}p^{20}=1+\frac{1}{2^{20}}+\frac{1}{3^{20}}+\frac{1}{4^{20}}+\frac{1}{5^{20}}+\text{etc.}\quad\text{etc.}$$

Of course this relationship which has been remarked upon has allowed me to proceed further than if I had used a direct method for finding the coefficients of the powers of p, where the work would certainly turn out to be extensive. Consequently I do not doubt that a distinguished service to the advancement of analysis will have been rendered when this extraordinary connection has been thoroughly investigated (for so far it is known to me only through observation). You will probably derive this connection without difficulty from the very nature of the thing.

While I have been writing these things, I have received from Nicolas Bernoulli, Professor of Law at Basle and a Fellow of your Society, a unique proof of the sum of this series

$$1 + \frac{1}{3^2} + \frac{1}{5^2} + \frac{1}{7^2} + \text{etc.}$$

which he has deduced from the sum of this known series

$$1 - \frac{1}{3} + \frac{1}{5} - \frac{1}{7} +$$

etc., in which he considers it as the square of the latter less twice the products of pairs of terms. [...]

But by writing such a long letter I fear that I may tax your forbearance in no small measure: wherefore I ask that you excuse my prolixity, and attribute it to the very high opinion of you which I formed long ago. [...]

(See Problem 28 for some exercises based upon this reading selection.)

Selection 3. Developing Rigor in Calculus

'Is Mathematical Truth Time-Dependent?'
Judith V. Grabiner
The American Mathematical Monthly, **81** (1974)
pages 354–365

1. Introduction

Is mathematical truth time-dependent? Our immediate impulse is to answer no. To be sure, we acknowledge that standards of truth in the natural sciences have undergone change; there was a Copernican revolution in astronomy, a Darwinian revolution in biology, an Einsteinian revolution in physics. But do scientific revolutions like these occur in mathematics? Mathematicians have most often answered this question as did the nineteenth-century mathematician Hermann Hankel, who said, "In most sciences, one generation tears down what another has built, and what one has established, the next undoes. In mathematics alone, each generation builds a new story to the old structure." [**20**, p. 25]

Hankel's view is not, however, completely valid. There have been several major upheavals in mathematics. For example, consider the axiomatization of geometry in ancient Greece, which transformed mathematics from an experimental science into a wholly intellectual one. Again, consider the discovery of non-Euclidean geometries and non-commutative algebras in the nineteenth century; these developments led to the realization that mathematics is not about anything in particular; it is instead the logically connected study of abstract systems. These were revolutions in thought which changed mathematicians' views about the nature of mathematical truth, and about what could or should be proved.

Another such mathematical revolution occurred between the eighteenth and nineteenth centuries, and was focussed primarily on the calculus. This change was a rejection of the mathematics of powerful techniques and novel results in favor of the mathematics of clear definitions and rigorous proofs. Because this change, however important it may have been for mathematicians themselves, is not often discussed by historians and philosophers, its revolutionary character is not widely understood. In this paper, I shall first try to show that this major change did occur. Then, I shall investigate what brought it about. Once we have done this, we can return to the question asked in the title of this paper.

2. Eighteenth-century Analysis: Practice and Theory

To establish what eighteenth-century mathematical practice was like, let us first look at a brilliant derivation of a now well-known result. Here is how Leonhard Euler derived the infinite series for the cosine of an angle. He began with the identity

$$(\cos z + i \sin z)^n = \cos nz + i \sin nz.$$

He then expanded the left-hand side of the equation according to the binomial theorem. Taking the real part of that binomial expansion and equating it to $\cos nz$, he obtained

$$\cos nz = (\cos z)^n - \frac{n(n-1)}{2!}(\cos z)^{n-2}(\sin z)^2 + \frac{n(n-1)(n-2)(n-3)}{4!}(\cos z)^{n-4}(\sin z)^4 - \cdots.$$

Let z be an infinitely small arc, and let n be infinitely large. Then:

$$\cos z = 1, \quad \sin z = z, \quad n(n-1) = n^2, \quad n(n-1)(n-2)(n-3) = n^4, \quad \text{etc.}$$

The equation now becomes recognizable:

$$\cos nz = 1 - \frac{n^2z^2}{2!} + \frac{n^4z^4}{4!} - \cdots.$$

But since z is infinitely small and n infinitely large, Euler concludes that nz is a finite quantity. So let $nz = v$. The modern reader may be left slightly breathless; still, we have

$$\cos v = 1 - \frac{v^2}{2!} + \frac{v^4}{4!} - \cdots.$$

(See [**16**, sections 133–4] and [**32**, pp. 348–9].)

Now that we have worked through one example, we shall be able to appreciate some generalizations about the way many eighteenth-century mathematicians worked. First, the primary emphasis was on getting results. All mathematicians know many of the results from this period, results which bear the names of Leibniz, Bernoulli, L'Hospital, Taylor, Euler, and Laplace. But the chances are good that these results were originally obtained in ways utterly different from the ways we prove them today. It is doubtful that Euler and his contemporaries would have been able to derive their results if they had been burdened with our standards of rigor. Here, then, is one major difference between the eighteenth-century way of doing mathematics and our way.

What led eighteenth-century mathematicians to think that results might be more important than rigorous proofs? One reason is that mathematics participated in the great explosion in science known as the Scientific Revolution [**19**]. Since the Renaissance, finding new knowledge had been a major goal of all the sciences. In mathematics, ever since the first major new result—the solution to the cubic equation published in 1545—increasing mathematical knowledge had meant finding new results. The invention of the calculus at the end of the seventeenth century intensified the drive for results; here was a powerful new method which promised vast new worlds to conquer. One can imagine few more exciting tasks than trying to solve the equations of motion for the whole solar system. The calculus was an ideal instrument for deriving new results, even though many mathematicians were unable to explain exactly why this instrument worked.

If the overriding goal of most eighteenth-century mathematics was to get results, we would expect mathematicians of the period to use those methods which produced results. For eighteenth-century mathematicians, the end justified the means. And the successes were many. New subjects arose in the eighteenth century, each with its own range of methods and its own domain of results: the calculus of variations, descriptive geometry, and partial differential equations, for instance. Also, much greater sophistication was achieved in existing subjects, like mathematical physics and probability theory.

The second generalization we shall make about eighteenth-century mathematics and its drive for results is that mathematicians placed great reliance on the power of symbols. Sometimes it seems to have been assumed that if one could just write down something which was symbolically coherent, the truth of the statement was guaranteed. And this assumption was not applied to finite formulas only. Finite methods were routinely extended to infinite processes. Many important facts about infinite power series were discovered by treating the series as very long polynomials [**30**].

This trust in symbolism in the eighteenth century is somewhat anomalous in the history of mathematics, and needs to be accounted for. It came both from the success of algebra and the success of the calculus. Let us first consider algebra. General symbolic notation of the type we now take for granted was introduced in 1591 by the French mathematician Francois Viète [**6**, pp. 59–65] and [**32**, pp. 74–81]. This notation proved to be the greatest instrument of discovery in the history of mathematics. Let us illustrate its power by one example. Consider the equation

$$(x-a)(x-b)(x-c) = x^3 - (a+b+c)\,x^2 + (ab+ac+bc)\,x - abc. \qquad (2.1)$$

Symbolic notation lets you discover what dozens of numerical examples may not: the relation between the roots and the coefficients of any polynomial equation of any degree. Equation (2.1), furthermore, has degree three, and has three roots. Relying on results like (2.1), Albert Girard in 1629 stated that an nth degree equation had n roots—the first formulation of what Gauss later called the Fundamental Theorem of Algebra.

But why are algebraic formulas like (2.1) considered true by eighteenth-century mathematicians? Because, as Newton put it, algebra is just a "universal arithmetic" [**29**]. Equation (2.1) is valid because it is a generalization about valid arithmetical statements. What, then, about infinite arguments, like the one of Euler's we examined earlier? The answer is analogous. Just as there is an arithmetic of infinite decimal fractions, we may generalize and create an algebra of infinite series [**28**, p. 6]. Infinite processes are like finite ones—except that they take longer.

The faith in symbolism nourished by algebra was enhanced further by the success of the calculus. Leibniz had invented the notations dy/dx and $\int y\,dx$ expressly to help us do our thinking. The notation serves this function well; we owe a debt to Leibniz every time we change variables under the integral sign. Or, suppose y is a function of x and that x is a function of t; we want to know dy/dt. It is not Leibniz, but Leibniz's notation that discovers the chain rule:

$$dy/dt = (dy/dx)(dx/dt).$$

The success of Leibniz's notation for the calculus reinforced mathematicians' belief in the power of symbolic arguments to give true conclusions.

In the eighteenth century, belief in the power of good notation extended beyond mathematics. For instance, it led the chemist Lavoisier to foresee a "chemical algebra," in the spirit of which Berzelius in 1813 devised chemical symbols essentially like those we use today. Anybody who has balanced chemical equations knows how the symbols do some of the thinking for us. The fact that the idea of the validity of purely symbolic arguments spread from mathematics to other areas shows us how prevalent an idea it must have been.

What has been said so far should not lead the reader to believe that eighteenth-century mathematicians were completely indifferent to the foundations of analysis. They certainly discussed the subject, and at length. I shall not here summarize the diverse eighteenth-century attempts to explain the nature of dy/dx, of limits, of the infinite, and of integrals, during a century that Carl Boyer has rightly called "the period of indecision" as far as foundations were concerned [**7**, Chapter VI]. What must be emphasized for our present purposes is that discussions of foundations were not the basic concern of eighteenth-century mathematicians. That is, discussions of foundations do not generally appear in research papers in scientific journals; instead, they are relegated to Chapter I of textbooks, or found in popularizations. More important, the practice of mathematics did not depend on a perfect understanding of the basic concepts used. But this was no longer the situation in nineteenth-century mathematics, and, of course, is not the situation today.

Nineteenth-century analysts, beginning with Cauchy and Bolzano, gave rigorous, inequality-based treatments of limit, convergence, and continuity, and demanded rigorous proofs of the theorems about these concepts. We know what these proofs were like; we still use them. This new direction in nineteenth-century analysis is not just a matter of differences in technique. It is a major change in the way mathematics was looked at and done. Now that we have sketched the eighteenth-century approach, we are ready to deal with what are—from the historical point of view—the most interesting questions of this paper. What made the change between the old and new views occur? How did mathematics get to be the way it is now?

Two things were necessary for the change. Most obviously, the techniques needed for rigorous proofs had to be developed. We shall discuss the history of some major techniques in Section 4, below. But also, there had to be a change in attitude. Without the techniques, of course, the change in attitude could never have borne fruit. But the change in attitude, though not sufficient, was a necessary condition for the establishment of rigor. Our next task, accordingly, will be to explain the change in attitude toward the foundations of the calculus between the eighteenth and nineteenth centuries. Did the very nature of mathematics force this change? Or was it motivated by factors outside of mathematics? Let us investigate various possibilities.

3. Why Did Standards of Mathematical Truth Change?

The first explanation which may occur to us is like the one we use to justify rigor to our students today: the calculus was made rigorous to avoid errors, and to correct errors already made. But this is not quite what happened. In fact, there are surprisingly few mistakes in eighteenth-century mathematics. There are two main reasons for this. First, some results could be verified numerically, or even experimentally; thus, their validity could be checked without a rigorous basis. Second, and even more important, eighteenth-century mathematicians had an almost unerring intuition. Though they were not guided by rigorous definitions, they nevertheless had a deep understanding of the properties of the basic concepts of analysis. This conclusion is supported by the fact that many apparently shaky eighteenth-century arguments can be salvaged, and made rigorous by properly specifying hypotheses. Nevertheless, we must point out that the need to avoid errors became more important near the end of the eighteenth century, when there was increasing interest among mathematicians in complex functions, in functions of several variables, and in trigonometric series. In these subjects, there are many plausible conjectures whose truth is relatively difficult to evaluate intuitively. Increased interest in such results may have helped draw attention to the question of foundations.

A second possible explanation which may occur to us is that the calculus was made rigorous in a spirit of generalization. The eighteenth century had produced a mass of results. The need to unify such a mass of results could have led automatically to a rigorous, axiomatic basis. But there had been large numbers of results for a hundred years before Cauchy's work. Besides, unifying results does not always make them rigorous; moreover, the function of rigor is not just to unify, but to prove. Still, there is something to be said for the hypothesis that the calculus became rigorous partly to unify the wealth of existing results. At the end of the eighteenth century, several mathematicians thought that the pace of getting new results was decreasing. This feeling had some basis in fact; most of the results obtainable by the routine application of eighteenth-century methods had been obtained. Perhaps, if progress was slowing, it was time to sit back and reflect about what had been done [**31**, pp. 136–7]. This feeling helped get some mathematicians interested in the question of rigor.

A third possible explanation depends on the prior existence of rigor in geometry. Everybody from the Greeks on knew that mathematics was supposed to be rigorous. One might thus assume that mathematicians' consciences began to trouble them, and that as a result analysts returned their new methods to the old standards. In fact, Euclidean geometry did provide a model for the new rigor. But the old ideas of rigor were not enough in themselves to make mathematicians strive to make the calculus rigorous—as the hundred and fifty years from Newton to Cauchy shows. This is true even though the discrepancy between Euclidean standards and the actual practice of eighteenth-century mathematicians did not go unnoticed. George Berkeley, Bishop of Cloyne, attacked the calculus in 1734, on the perfectly valid grounds that it was not rigorous the way mathematics was supposed to be. Berkeley wanted to defend religion against the attacks of unreasonableness levelled against it by eighteenth-century scientists and mathematicians. Berkeley said that his opponents did not even reason well about mathematics. He conceded that the results of the calculus were valid, but attacked its methods. Berkeley's attack, *The Analyst*, is a masterpiece of polemics [**32**, pp. 333–338] and [**3**]. He said of the "vanishing increments" that played so crucial a role in Newton's calculus, "And what are these... vanishing increments? They are neither finite quantities, nor quantities infinitely small, nor yet nothing. May we not call them the ghosts of departed quantities?" Berkeley's attack—which included point-by-point mathematical criticisms of some basic arguments of Newton's calculus—provoked a number of mathematicians to write refutations. However, neither Berkeley's attack nor the replies to it produced the change in attitude toward rigor which we are trying to explain. First of all, the replies are not very convincing [**8**]. Besides, the subject of foundations was still not considered serious mathematics. Berkeley did get people thinking, more than they would have without him, about the problem of foundations. The discussions of foundations by Maclaurin, D'Alembert, and Lagrange were all at least somewhat influenced by Berkeley's work. Nevertheless, Berkeley's attack in itself was not enough to cause foundations to become a major mathematical concern.

In bringing about the change, there is one other factor which, though seldom mentioned in this connection, was important: the mathematician's need to teach. Near the end of the eighteenth century, a major social change occurred. Before the last decades of the century, mathematicians were often attached to royal courts; their job was to do mathematics and thus add to the glory, or edification, of their patron. But almost all mathematicians since the French Revolution have made their living by teaching [**31**, p. 140] [**2**, p. 95,108].

This change in the economic circumstances of mathematicians had other causes than the decline of particular royal courts. In the eighteenth century, science was expanding. This was the "age of Newton" and the success of Newtonian science. Governments and businessmen felt that science was important and could be useful; scientists encouraged them in these beliefs. So governments founded educational institutions to promote science. Military schools were founded to provide prospective officers with knowledge of applied science. New scientific chairs were endowed in existing universities. By far the most important new institution for scientific instruction, one which served as a model to several nations in the nineteenth century, was the *École polytechnique* in Paris, founded in 1795 by the revolutionary government in France.

Why might the new economic circumstances of mathematicians—the need to teach—have helped promote rigor? Teaching always makes the teacher think carefully about the basis for the subject. A mathematician could understand enough about a concept to use it, and could rely on the insight he had gained through his experience. But this does not work with freshmen, even in the eighteenth century. Beginners will not accept being told, "After you have worked with this concept for three years, you'll understand it."

What is the evidence that teaching helped motivate eighteenth and nineteenth century mathematicians to make analysis rigorous? First, until the end of the eighteenth century, most work on foundations did not appear in scientific journals, apparently because foundations were not considered to pose major mathematical (as opposed to philosophical) questions. Instead, such work appeared in courses of lectures, in textbooks, or in popularizations. Even in the nineteenth century, when foundations had been established as essential to mathematics, their origin was often in teaching. The work on foundations of analysis of Lagrange [**23**, **26**], of Cauchy [**10**, **11**], of Weierstrass [**21**, pp. 283–4] [**7**, pp. 284–7], and of Dedekind [**14**, p. 1], all originated in courses of lectures.

Each of the points we have made so far helps explain what motivated mathematicians to shift from the result-oriented view of the eighteenth century to the more rigorous standards of the nineteenth. One more catalyst of the change should be identified: Joseph-Louis Lagrange. Lagrange's own interest in the problem of foundations was first engaged by having to teach the calculus at the military school in Turin [**24**]. In 1784, by proposing the foundations of the calculus as a prize problem for the Berlin Academy of Sciences, he stimulated the first major book-length contributions to foundations of the calculus written on the Continent. (see [**27**] [**9**] [**7**, pp. 254–255] and [**18**, pp. 149–150]). Above all, Lagrange's lectures at the *École polytechnique*, published in two widely influential books, attempted to give a general and algebraic framework for the calculus [**26**] [**23**]. Lagrange did not correctly solve the problem of foundations—we can no longer accept his *definition* of $f'(x)$ as the coefficient of h in the Taylor series expansion of $f(x+h)$. Nevertheless, his vision of reducing the calculus to algebra decisively influenced the work of Bolzano [**5**] and—as we shall see—of Cauchy.

The change in attitude we have been discussing was not enough in itself to establish rigor in the calculus—as the example of Lagrange shows. Having decided that we want to make a subject rigorous, what else do we need? Two more things are required: the right definitions, and techniques of proof to derive the known results from the definitions. We must now answer another question: where did the required definitions and proofs come from?

Eighteenth-century mathematicians themselves had developed many of the techniques, and isolated many of the basic defining properties—even though they did not know that this is what they were doing. It is amazing that so many of the techniques used by Cauchy in rigorous arguments had been around for so long. This fact shows that a real change in point of view was required for the rigorization of analysis; it was not an automatic development out of eighteenth-century mathematics.

4. The Eighteenth-century Origins of Nineteenth-century Rigor

We shall illustrate the eighteenth-century origins of nineteenth-century rigor by giving several examples of eighteenth-century work which was transformed into nineteenth-century definitions and proofs. The principle area of eighteenth-century mathematics we shall investigate is the study of approximations. Eighteenth-century mathematicians, whether solving algebraic equations or differential equations, developed many useful approximation methods. When the goal is results, an approximate result is better than nothing. Paradoxically, eighteenth-century mathematicians were most exact when they were being approximate; their work with inequalities in approximations later became the basis for rigorous analysis.

We shall discuss two classes of eighteenth-century approximation work: the actual working out of approximation procedures, and computation of error estimates. Let us see what use nineteenth-century analysts made of these.

One new way in which nineteenth-century mathematicians looked at eighteenth-century approximations was to see the approximate solution as a construction of that solution, and therefore as a proof of its existence. For instance, Cauchy did this in developing what is now called the Cauchy-Lipschitz method of proving the existence of the solution to a differential equation; the proof is based on an approximation method developed by Euler [**15**, pp. 424–5; **12**, p. 399ff]. Similarly Cauchy's elegant proof of the intermediate-value theorem for continuous functions was based on an eighteenth-century approximation method [**22**, pp. 260–1; **25**, sections 2,6; **10**, pp. 378–80]. For a continuous function $f(x)$, Cauchy took $f(a)$ and $f(b)$ of opposite sign, divided the interval $[a, b]$ into n parts, and concluded that there were at least two values of x on $[a, b]$, differing by $(b - a)/n$, which yielded opposite sign for $f(x)$. He then repeated the procedure on the interval between these two new values, on an interval of length $(b - a)/n$, which gives two more values, differing by $(b - a)/n^2$, and so on. Where Lagrange had used this technique to approximate the root ξ of a polynomial included between $x = a$ and $x = b$, Cauchy used it to argue for the existence of the number ξ as the common limit of the sequences of values of x which gave positive sign for f, and negative sign for f. The origin of Cauchy's proof in algebraic approximations is further demonstrated by the context in which he gave it: a "*Note*" devoted to discussing the approximate solution of algebraic equations [**10**, p. 378 ff].

Another example of the conversion of approximations into existence proofs is given by Cauchy's theory of the definite integral. In the eighteenth century, it was customary to define the integral as the inverse of the derivative. It was known, however, that the value of the integral could be approximated by a sum. Cauchy took Euler's work on approximating the values of definite integrals by sums [**15**, pp. 184–7], and looked at it from an entirely new point of view. Cauchy *defined* the definite integral as the limit of a sum, proved the existence of the definite integral of a continuous (actually, uniformly continuous) function, and then used his definition to prove the Fundamental Theorem of Calculus [**11**, pp. 122–5, 151–2].

Now let us consider another type of result in eighteenth-century approximations: approximations given along with an error estimate. These results took a form like this: given some n, the mathematician could compute an upper bound on the error made in taking the nth approximation for the true value. Near the end of the eighteenth century, the algebra of inequalities was exploited with great skill in computing such error estimates [**13**, pp. 171–183] and [**25**, pp. 46–7, p. 163]. Cauchy, Abel, and their followers turned the approximating process around. Instead of being given n and finding the greatest possible error, we are *given* what is in effect the "error"—epsilon—and, provided that the process converges, we can always find n such that the error of the nth approximation is less than epsilon. (This seems to be the reason for the use of the letter "epsilon" in its usual modern sense by Cauchy [**10**, pp. 64–5 *et passim*].) [**1**] [**10**, pp. 400–415]. Cauchy's definition of convergence—which is essentially ours—is based on this principle [**10**, Chapter VI].

Another way in which nineteenth-century mathematicians changed eighteenth-century views of results using inequalities was to take facts known to eighteenth-century mathematicians in special cases and to make them legitimate in general. For instance, D'Alembert and others had shown that some particular series converged by showing that they were, term-by-term, less than a convergent geometric progression [**13**]. Gauss in 1813 used this criterion to investigate, in a rigorous manner, the convergence of the hypergeometric series [**17**]. Cauchy used the comparison of a given series with a geometric one to derive and to prove some general tests for the convergence of any series; the ratio test, the logarithm test, and the root test [**10**, pp. 121–127].

Let us look at one last example—a very important one—of an eighteenth-century result which became something different in the nineteenth century: the property of the derivative expressed by

$$f(x+h) = f(x) + hf'(x) + hV, \tag{4.1}$$

where V goes to zero with h. As we have remarked, Lagrange had defined $f'(x)$ as the coefficient of h in the Taylor expansion of $f(x+h)$. He then "derived" (4.1) from that Taylor series expansion, considering V to be a convergent infinite series in h. Lagrange used (4.1) to investigate many properties of the derivative. To do this, he interpreted "V goes to zero with h" to mean that, for any given quantity D, we can find h sufficiently small so that $f(x+h) - f(x)$ "will be included between" $h[f'(x) - D]$ and $h[f'(x) + D]$ [**23**, p. 87]. First Cauchy, and then Bolzano and Weierstrass, made (4.1) and its associated inequalities into the *definition* of $f'(x)$. (Cauchy's definition was actually verbal, but he translated it into the language of inequalities in proofs.) [**11**, pp. 44–5; 122–3], [**4**, Chapter 2] and [**7**, pp. 285–7]. This definition made legitimate the results about $f'(x)$ that Lagrange had derived from (4.1)—for instance, the mean-value theorem for derivatives. (Except, we must note, for a few errors, especially the confusion between convergence and uniform convergence, which was not cleared up until the 1840's.)

Of course, we do not mean to imply that Gauss, Cauchy, Bolzano, Abel, and Weierstrass were not original, creative mathematicians. They were. To show that major changes in point of view occur in mathematics, we have concentrated in this section on what these men owed to eighteenth-century techniques. But, besides transforming what they borrowed, they contributed much of their own that was new. Cauchy, in particular, devised beautiful proofs about convergent power series in real and complex variables, about real and complex integrals, and, of course, contributed to a variety of subjects besides analysis. Nevertheless, for our present purposes, we need the biassed sample we have chosen—things accomplished either by taking what the eighteenth century knew for particular cases and making it general, or by taking what the eighteenth century had derived for one purpose and putting it to a more profound use.

Much effort was needed to transform eighteenth-century techniques in the ways we have discussed. But it was more than just a matter of effort. It took asking the right questions *first*; and then using—and expanding—the already existing techniques to answer them. It took—and was—a major change in point of view. The reawakening of interest in rigor was just as necessary as the availability of techniques to produce the point of view of Bolzano and Cauchy—the point of view which has been with us ever since. Mathematics requires not only results, but clear definitions and rigorous proofs. Individual mathematicians may still concentrate on the creation of fruitful methods and ideas to be exploited, but the mathematical community as a whole can no longer be indifferent to rigor.

5. Conclusion

We began by asking whether mathematical truth was time-dependent. Perhaps mathematical truth is eternal, but our knowledge of it is not. We have now seen an example of how attitudes toward mathematical truth have changed in time. After such a revolution in thought, earlier work is re-evaluated. Some is considered worth more; some, worth less.

What should a mathematician do, knowing that such re-evaluations occur?

Three courses of action suggest themselves. First, we can adopt a sort of relativism which has been expressed in the phrase "Sufficient unto the day is the rigor thereof." Mathematical truth is just what the editors of the *Transactions* say it is. This is a useful view at times. But this view, if universally adopted, would mean that Cauchy and Weierstrass would never have come along. Unless there were the prior appearance of major errors, standards could never improve in any important way. So the attitude of relativism, which would have counselled Cauchy to leave foundations alone, will not suffice for us.

Second, we can attempt to set the highest conceivable standard: never use an argument in which we do not completely understand what is going on, dotting all the i's and crossing all the t's. But this is even worse. Euler, after all, knew that there were problems in dealing with infinitely large and infinitely small quantities. Acording to this high standard, which textbooks sometimes urge on students, Euler would never have written a line. There would have been no mathematical structure for Cauchy and Weierstrass to make rigorous.

So I suggest a third possibility: a recognition that the problem I have raised is just the existential situation mathematicians find themselves in. Mathematics grows in two ways: not only by successive increments, but also by occasional revolutions. Only if we accept the possibility of present error can we hope that the future will bring a fundamental improvement in our knowledge. We can be consoled that most of the old bricks will find places somewhere in the new structure. Mathematics is *not* the unique science without revolutions. Rather, mathematics is that area of human activity which has at once the least destructive and still the most fundamental revolutions.

This paper was originally delivered at the Mathematical Association of America, Southern California Section, March 1972. The author wishes to thank Elmer Tolsted for encouragement and suggestions.

References

1. N. H. Abel, Recherches sur la série $1 + (m/1)x + [m(m-1)/1.2]x^2 + [m(m-1)(m-2)/1.2.3]x^3 + \cdots$, Oeuvres complètes, Vol. I, Christiania, 1881.
2. J. Ben David, The Scientist's Role in Society, Prentice-Hall, Englewood Cliffs, 1971.
3. G. Berkeley, The Works of George Berkeley, Vol. IV, ed. A.A. Luce and T.E. Jessop, Edinburgh, 1948–1957.
4. B. Bolzano, Functionenlehre, Schriften, Band I, Prague, 1930.

5. ———, Rein analytischer Beweis des Lehrsatzes, dass zwischen je zwei Werthen, die ein entgegengesetztes Resultat gewaehren, wenigstens eine reelle Wurzel der Gleichung liege, 1817, Englemann, Leipzig, 1905.
6. Carl Boyer, History of Analytic Geometry, Scripta Mathematica, New York, 1956.
7. ———, History of the Calculus and its Conceptual Development, Dover, New York, 1959.
8. F. Cajori, A History of the Conceptions of Limits and Fluxions in Great Britain from Newton to Woodhouse, Open Court, Chicago, 1931.
9. L.N.M. Carnot, Réflexions sur la métaphysique de calcul infinitésimal, Duprat, Paris, 1797.
10. A.-L. Cauchy, Cours d'analyse de l'école royale polytechnique, Imprimerie royale, Paris, 1821, in Oeuvres Complètes, Series 2, Vol. III, Gauthier-Villars, Paris, 1897.
11. ———, Résumé des lecons données à l'école royale polytechique sur le calcul infinitésimal, Imprimerie royale, Paris, 1823, in Oeuvres Complètes, Series 2, Vol. IV, Gauthier-Villars, Paris, 1899.
12. ———, Exercices d'analyse, 1840, in Oeuvres, Series 2, Vol. XI.
13. Jean D'Alembert, Réflexions sur les suites et sur les racines imaginaires, Opuscules mathématiques, vol. V, Paris 1768, pp. 171–215.
14. Richard Dedekind, Essays on the theory of numbers, Dover, New York, 1963.
15. Leonhard Euler, Institutiones calculi integralis 1768, Opera Omnia, Series 1, vol. XI, Teubner, Leibniz and Berlin, 1911.
16. ———, Introductio in analysin infinitorum 1748, Opera Omnia, Series 1, vols, 8–9.
17. K.F. Gauss, Disquisitio generales circa seriem infinitam

$$1+\frac{\alpha\cdot\beta}{1\cdot\gamma}x+\frac{\alpha(\alpha+1)\beta(\beta+1)}{1\cdot 2\cdot\gamma(\gamma+1)}x^2+\frac{\alpha(\alpha+1)(\alpha+2)\beta(\beta+1)(\beta+2)}{1\cdot 2\cdot 3\cdot\gamma(\gamma+1)(\gamma+2)}x^3+\cdots, [1813],$$

Werke, Vol. 3, pp. 123–162; German translation, Berlin, 1888.
18. C.C. Gillispie, Lazare Carnot Savant, Princeton, 1971.
19. A.R. Hall, The Scientific Revolution, 1500–1800. Beacon, Boston, 1966.
20. H. Hankel, Die Entwicklung der Mathematik im letzten Jahrhundert, 1884, quoted by M.Moritz, On Mathematics and Mathematicians, Dover, New York, 1942, p. 14.
21. F. Klein, Vorlesungen über die Entwicklung der Mathematik im 19. Jahrhundert, 1926, reprinted by Chelsea, New York, 1967.
22. J.-L. Lagrange, Lecons élémentaires sur les mathématiques, données à l' école normale en 1795, Oeuvres, VIII, Gauthier-Villars, Paris, 1867–1892, pp. 181–288.
23. ———, Lecons sur le calcul des fonctions, 2d edition, 1806, Oeuvres, X.
24. ———, Letter to Euler, 24 November 1759, Oeuvres, XIV, pp. 170–174.
25. ———, Traité de la résolution des équations numériques de tous les degrés, 1808, Oeuvres,, VIII.
26. ———, Théorie des fonctions analytiques, 2d. edition, 1813, Oeuvres, IX.
27. S. L'Huilier, Exposition élémentaire des principes des calculs supérieurs, Decker, Berlin, 1787.
28. Isaac Newton, On the analysis by equations of an infinite number of terms, 1669, in D.T. Whiteside, ed., The Mathematical Works of Isaac Newton, Johnson Reprint, London and New York, 1964, vol. I.

29. ———, Universal Arithmetic, 1707, in D.T. Whiteside ed., The Mathematical Works of Isaac Newton, vol II, Johnson, London and New York, 1970.
30. R. Reiff, Geschichte der unendlichen Reihen, Tübingen, 1889.
31. D.J. Struik, Concise History of Mathematics, Dover, New York, 1967.
32. ———, ed., A Source Book in Mathematics, 1200–1800, Harvard, Cambridge, 1967.

SMALL COLLEGE, CALIFORNIA STATE COLLEGE, DOMINIGUES HILLS, CA 90505

Selection 4. The Contribution of Lebesgue

From *Biographical Dictionary of Mathematicians*
by Charles Coulston Gillispie, Editor-in-Chief
Charles Scribner's Sons, N.Y., 1995
pages 1470–1472
Reprinted by permission of The Gale Group

LEBESGUE, HENRI LÉON (*b.* Beauvais, France, 28 June 1875; *d.* Paris, France, 26 July 1941)

Lebesgue studied at the École Normale Supérieure from 1894 to 1897. His first university positions were at Rennes (1902–1906) and Poitiers (1906–1910). At the Sorbonne, he became *maître de conférences* (in mathematical analysis, 1910–1919) and then *professeur d'application de la géométrie à l'analyse*. In 1921 he was named professor at the Collège de France and the following year was elected to the Académie des Sciences.

Lebesgue's outstanding contribution to mathematics was the theory of integration that bears his name and that became the foundation for subsequent work in integration theory and its applications. After completing his studies at the École Normale Supérieure, Lebesgue spent the next two years working in its library, where he became acquainted with the work of another recent graduate, René Baire. Baire's unprecedented and successful researches on the theory of discontinuous functions of a real variable indicated to Lebesgue what could be achieved in this area. From 1899 to 1902, while teaching at the Lycée Centrale in Nancy, Lebesgue developed the ideas that he presented in 1902 as his doctoral thesis at the Sorbonne. In this work Lebesgue began to develop his theory of integration which, as he showed, includes within its scope all the bounded discontinuous functions introduced by Baire.

The Lebesgue integral is a generalization of the integral introduced by Riemann in 1854. As Riemann's theory of integration was developed during the 1870's and 1880's, a measure-theoretic viewpoint was gradually introduced. This viewpoint was made especially prominent in Camille Jordan's treatment of the Riemann integral in his *Cours d'analyse* (1893) and strongly influenced Lebesgue's outlook on these matters. The significance of placing integration theory within a measure-theoretic context was that it made it possible to see that a generalization of the notions of the measure and measurability carries with it corresponding generalizations of the notions of the integral and integrability. In 1898, Émile Borel was led through his work on complex function theory to introduce radically different notions of measure and measurability. Some mathematicians found Borel's ideas lacking in appeal and relevance, especially since they involved assigning measure zero to some dense sets. Lebesgue, however, accepted them. He completed Borel's definitions of measure and measurability so that they represented generalizations of Jordan's definitions and then used them to obtain his generalization of the Riemann integral.

After the work of Jordan and Borel, Lebesgue's generalizations were somewhat inevitable. Thus, W. H. Young and G. Vitali, independently of Lebesgue and of each other, introduced the same generalization of Jordan's theory of measure; in Young's case, it led to a generalization of the integral that was essentially the same as Lebesgue's. In Lebesgue's work, however, the generalized definition of the integral was simply the starting point of his contributions to integration theory. What made the new definition important was that Lebesgue was able to recognize in it an analytical tool capable of dealing with—and to a large extent overcoming—the numerous theoretical difficulties that had arisen in connection with Riemann's theory of integration. In fact, the problems posed by these difficulties motivated all of Lebesgue's major results.

The first such problem had been raised unwittingly by Fourier early in the nineteenth century: (1) If a bounded function can be represented by a trigonometric series, is that series the Fourier series of the function? Closely related to (1) is (2): When is the term-by-term integration of an infinite series permissible? Fourier had assumed that for bounded functions that answer to (2) is Always, and he had used this assumption to prove that the answer to (1) is Yes.

By the end of the nineteenth century it was recognized—and emphasized—that term-by-term integration is not always valid even for uniformly bounded series of Riemann-integrable functions, precisely because the function represented by the series need not be Riemann-integrable. These developments, however, paved the way for Lebesgue's elegant proof that term-by-term integration is permissible for any uniformly bounded series of Lebesgue-integrable functions. By applying this result to (1), Lebesgue was able to affirm Fourier's conclusion that the answer is Yes.

Another source of difficulties was the fundamental theorem of the calculus,

$$\int_a^b f'(x)\,dx = f(b) - f(a).$$

The work of Dini and Volterra in the period 1878–1881 made it clear that functions exist which have bounded derivatives that are not integrable in Riemann's sense, so that the fundamental theorem becomes meaningless for these functions. Later further classes of these functions were discovered; and additional problems arose in connection with Harnack's extension of the Riemann integral to unbounded functions because continuous functions with densely distributed intervals of invariability were discovered. These functions provided examples of Harnack-integrable derivatives for which the fundamental theorem is false. Lebesgue showed that for bounded derivatives these difficulties disappear entirely when integrals are taken in his sense. He also showed that the fundamental theorem is true for an unbounded, finite-valued derivative f' that is Lebesgue-integrable and that this is the case if, and only if, f is of bounded variation. Furthermore, Lebesgue's suggestive observations concerning the case in which f' is finite-valued but not Lebesgue-integrable were successfully developed by Arnaud Denjoy, starting in 1912, using the transfinite methods developed by Baire.

The discovery of continuous monotonic functions with densely distributed intervals of invariability also raised the question: When is a continuous function an integral? The question prompted Harnack to introduce the property that has since been termed absolute continuity. During the 1890's absolute continuity came to be regarded as the characteristic property of absolutely convergent integrals, although no one was actually able to show that every absolutely continuous function is an integral. Lebesgue, however, perceived that this is precisely the case when integrals are taken in his sense.

A deeper familiarity with infinite sets of points had led to the discovery of the problems connected with the fundamental theorem. The nascent theory of infinite sets also stimulated an interest in the meaningfulness of the customary formula

$$L = \int_a^b [1 + (f')^2]^{1/2}$$

for the length of the curve $y = f(x)$. Paul du Bois-Reymond, who initiated an interest in the problem in 1879, was convinced that the theory of integration is indispensable for the treatment of the concepts of rectifiability and curve length within the general context of the modern function concept. But by the end of the nineteenth century this view appeared untenable, particularly because of the criticism and counterexamples given by Ludwig Scheeffer. Lebesgue was quite interested in this matter and was able to use the methods and results of his theory of integration to reinstate the credibility of du Bois-Reymond's assertion that the concepts of curve length and integral are closely related.

Lebesgue's work on the fundamental theorem and on the theory of curve rectification played an important role in his discovery that a continuous function of bounded variation possesses a finite derivative except possibly on a set of Lebesgue measure zero. This theorem gains in significance when viewed against the background of the century-long discussion of the differentiability properties of continuous functions. During roughly the first half of the nineteenth century, it was generally thought that continuous functions are differentiable at "most" points, although continuous functions were frequently assumed to be "piecewise" monotonic. (Thus, differentiability and monotonicity were linked together, albeit tenuously.) By the end of the century this view was discredited, and no less a mathematician than Weierstrass felt that there must exist continuous monotonic functions that are nowhere differentiable. Thus, in a sense, Lebesgue's theorem substantiated the intuitions of an earlier generation of mathematicians.

Riemann's definition of the integral also raised problems in connection with the traditional theorem positing the identity of double and iterated integrals of a function of two variables. Examples were discovered for which the theorem fails to hold. As a result, the traditional formulation of the theorem had to be modified, and the modifications became drastic when Riemann's definition was extended to unbounded functions. Although Lebesgue himself did not resolve this infelicity, it was his treatment of the problem that formed the foundation for Fubini's proof (1907) that the Lebesgue integral does make it possible to restore to the theorem the simplicity of its traditional formulation.

During the academic years 1902–1903 and 1904–1905, Lebesgue was given the honor of presenting the Cours Peccot at the Collège de France. His lectures, published as the monographs *Lecons sur l'intégration* . . . (1904) and *Lecons sur les séries trigonométriques* (1906), served to make his ideas better known. By 1910 the number of mathematicians engaged in work involving the Lebesgue integral began to increase rapidly. Lebesgue's own work—particularly his highly successful applications of his integral in the theory of trigonometric series—was the chief reason for this increase, but the pioneering research of others, notably Fatou, F. Rietz, and Fischer, also contributed substantially to the trend. In particular, Riesz's work on L^p spaces secured a permanent place for the Lebesgue integral in the theory of integral equations and function spaces.

Although Lebesgue was primarily concerned with his own theory of integration, he played a role in bringing about the development of the abstract theories of measure and integration that predominate in contemporary mathematical research. In 1910 he published an important memoir, "Sur l'intégration des fonctions discontinues," in which he extended the theory of integration and differentiation to n-dimensional space. Here Lebesgue introduced, and made fundamental, the notion of a countably additive set function (defined on Lebesgue-measurable sets) and observed in passing that such functions are of bounded variation on sets on which they take a finite value. By thus linking the notions of bounded variation and additivity, Lebesgue's observation suggested to Radon a definition of the integral that would include both the definitions of Lebesgue and Stieltjes as special cases. Radon's paper (1913), which soon led to further abstractions, indicated the viability of Lebesgue's ideas in a much more general setting.

By the time of his election to the Académie des Sciences in 1922, Lebesgue had produced nearly ninety books and papers. Much of this output was concerned with his theory of integration, but he also did significant work on the structure of sets and functions (later carried further by Lusin and others), the calculus of variations, the theory of surface area, and dimension theory.

For his contributions to mathematics Lebesgue received the Prix Houllevigue (1912), the Prix Poncelet (1914), the Prix Saintour (1917), and the Prix Petit d'Ormoy (1919). During the last twenty years of his life, he remained very active, but his writings reflected a broadening of interests and were largely concerned with pedagogical and historical questions and with elementary geometry.

BIOGRAPHICAL INFORMATION AND REFERENCES ON LEBESGUE CAN BE OBTAINED FROM K. O. MAY, "BIOGRAPHICAL SKETCH OF HENRI LEBESGUE," IN H. LEBESGUE, *Measure and the Integral*, K. O. MAY, ED., (SAN FRANCISCO, 1966), 1–7. FOR A DISCUSSION OF LEBESGUE'S WORK ON INTEGRATION THEORY AND ITS HISTORICAL BACKGROUND, SEE T. HAWKINS, *Lebesgue's Theory of Integration: Its Origins and Development* (MADISON, WIS., 1970).

THOMAS HAWKINS

Selection 5. The Bernoulli Numbers and Some Wonderful Discoveries of Euler

Calculus Gems
George F. Simmons
McGraw-Hill Inc., 1992
pages 298–302
The following is reproduced with permission of the McGraw-Hill Companies.

In this section we derive several formulas discovered by Euler that rank among the most elegant truths in the whole of mathematics. We use the word "derive" instead of "prove" because some of our arguments are rather formal and require more advanced ideas than we can provide here to become fully rigorous in the sense demanded by modern concepts of mathematical proof. However, the mere fact that we are not able to seal every crack in the reasoning seems a flimsy excuse for denying students an opportunity to glimpse some of the wonders that can be found in this part of calculus. For those who wish to dig deeper, full proofs are given in the treatise by K. Knopp, *Theory and Application of Infinite Series*, Hafner, 1951.

The Bernoulli Numbers

Since

$$\frac{e^x - 1}{x} = 1 + \frac{x}{2!} + \frac{x^2}{3!} + \cdots$$

for $x \neq 0$, and this power series has the value 1 at $x = 0$, the reciprocal function $x/(e^x - 1)$ has a power series expansion valid in some neighbourhood of the origin if the value of this function is defined to be 1 at $x = 0$. We write this series in the form

$$\frac{x}{e^x - 1} = \sum_{n=0}^{\infty} \frac{B_n}{n!} x^n = B_0 + B_1 x + \frac{B_2}{2!} x^2 + \dots. \tag{1}$$

The numbers B_n defined in this way are called the *Bernoulli numbers*, and it is clear that $B_0 = 1$. A bit of algebra reveals that

$$\frac{x}{e^x - 1} = \frac{x}{2}\left(\frac{e^x + 1}{e^x - 1} - 1\right) = -\frac{x}{2} + \frac{x}{2} \cdot \frac{e^x + 1}{e^x - 1}. \tag{2}$$

A routine check shows that the second term on the right is an even function, so $B_1 = -\frac{1}{2}$ and $B_n = 0$ if n is odd and > 1. If we write (1) in the form

$$\left(\frac{B_0}{0!} + \frac{B_1}{1!} x + \frac{B_2}{2!} x^2 + \cdots\right)\left(\frac{1}{1!} + \frac{x}{2!} + \frac{x^2}{3!} + \cdots\right) = 1,$$

then it is clear that the coefficient of x^{n-1} in the product on the left equals zero if $n > 1$. By the rule for multiplying power series, this yields

$$\frac{B_0}{0!}\cdot\frac{1}{n!}+\frac{B_1}{1!}\cdot\frac{1}{(n-1)!}+\frac{B_2}{2!}\cdot\frac{1}{(n-2)!}+\cdots+\frac{B_{n-1}}{(n-1)!}\cdot\frac{1}{1!}=0,$$

and by multiplying through by $n!$ we obtain

$$\frac{n!}{0!n!}B_0+\frac{n!}{1!(n-1)!}B_1+\frac{n!}{2!(n-2)!}B_2+\cdots+\frac{n!}{(n-1)!1!}B_{n-1}=0. \tag{3}$$

This equation can also be written more briefly as

$$\binom{n}{0}B_0+\binom{n}{1}B_1+\binom{n}{2}B_2+\cdots+\binom{n}{n-1}B_{n-1}=0$$

or

$$\sum_{k=0}^{n-1}\binom{n}{k}B_k=0,$$

where $\binom{n}{k}$ is the binomial coefficient $n!/[k!(n-k)!]$. By taking $n = 3, 5, 7, 9, 11, \ldots$ in (3) and doing a little arithmetic, we find that

$$B_2=\frac{1}{6},\quad B_4=-\frac{1}{30},\quad B_6=\frac{1}{42},\quad B_8=-\frac{1}{30},\quad B_{10}=\frac{5}{66},\ldots$$

These calculations can be continued recursively as far as we please, so all the Bernoulli numbers can be considered as known, even though considerable labor may be required to make any particular one of them visibly present. We also point out that it is obvious from (3) and the mode of calculation that every B_n is rational.

The Power Series for the Tangent

We now begin to explore the uses of these numbers.

In equation (2) we move the term $-x/2$ to the left and use the fact that

$$\frac{x}{2}\cdot\frac{e^x+1}{e^x-1}=\frac{x}{2}\cdot\frac{e^{x/2}+e^{-x/2}}{e^{x/2}-e^{-x/2}}$$

to obtain

$$\frac{x}{2}\cdot\frac{e^{x/2}+e^{-x/2}}{e^{x/2}-e^{-x/2}}=\sum_{n=0}^{\infty}\frac{B_{2n}}{(2n)!}x^{2n}, \tag{4}$$

On the left-hand side of this, we now replace x by $2ix$, which yields

$$\frac{2ix}{2}\cdot\frac{e^{ix}+e^{-ix}}{e^{ix}-e^{-ix}}=x\frac{(e^{ix}+e^{-ix})/2}{(e^{ix}-e^{-ix})/2i}=x\cot x,$$

by means of the well-known formulas expressing $\sin x$ and $\cos x$ in terms of the exponential function. Making the same substitution on the right-hand side of (4) gives

$$\sum_{n=0}^{\infty} \frac{B_{2n}}{(2n)!}(2ix)^{2n} = \sum_{n=0}^{\infty} (-1)^n \frac{2^{2n} B_{2n}}{(2n)!} x^{2n},$$

so

$$x \cot x = \sum_{n=0}^{\infty} (-1)^n \frac{2^{2n} B_{2n}}{(2n)!} x^{2n}. \tag{5}$$

The trigonometric identity $\tan x = \cot x - 2 \cot 2x$ now enables us to use (5) to write

$$\begin{aligned}
\tan x &= \sum_{n=0}^{\infty} (-1)^n \frac{2^{2n} B_{2n}}{(2n)!} x^{2n-1} - 2 \sum_{n=0}^{\infty} (-1)^n \frac{2^{2n} B_{2n}}{(2n)!} (2x)^{2n-1} \\
&= \sum_{n=0}^{\infty} (-1)^n \frac{2^{2n} B_{2n}}{(2n)!} x^{2n-1} - \sum_{n=0}^{\infty} (-1)^n \frac{2^{2n} B_{2n}}{(2n)!} 2^{2n} x^{2n-1} \\
&= \sum_{n=0}^{\infty} (-1)^n \frac{2^{2n} B_{2n}}{(2n)!} (1 - 2^{2n}) x^{2n-1},
\end{aligned}$$

so

$$\tan x = \sum_{n=1}^{\infty} (-1)^{n+1} \frac{2^{2n}(2^{2n} - 1) B_{2n}}{(2n)!} x^{2n-1}.$$

This is the full power series for $\tan x$ that is usually encountered several times in truncated form in elementary treatments of infinite series. Based on our knowledge of the Bernoulli numbers, the first few terms of this series are easy to calculate explicitly,

$$\tan x = x + \frac{1}{3}x^3 + \frac{2}{15}x^5 + \frac{17}{315}x^7 + \frac{67}{2835}x^9 + \cdots.$$

The Partial Fractions Expansion of the Cotangent

By using entirely different methods, Euler discovered another remarkable expansion of the cotangent: If x is not an integer, then

$$\pi \cot \pi x = \frac{1}{x} + 2x \sum_{n=1}^{\infty} \frac{1}{x^2 - n^2}. \tag{6}$$

We will examine this formula from two very different points of view, and give two derivations.

First, it is quite easy to see that (6) is analogous to the expansion of a rational function in partial fractions. For instance, if we consider the rational function

$$\frac{2x+1}{x^2 - 3x + 2}$$

and notice that the denominator has zeros 1 and 2 and can therefore be factored into $(x-1)(x-2)$, then this leads to the expansion

$$\frac{2x+1}{x^2-3x+2} = \frac{2x+1}{(x-1)(x-2)} = \frac{c_1}{x-1} + \frac{c_2}{x-2}$$

for certain constants c_1 and c_2. The constant c_1 can now be determined by multiplying through by $x-1$ and allowing x to approach 1, and similarly for c_2. Formally, (6) can be obtained in much the same way by noticing that

$$\cot \pi x = \frac{\cos \pi x}{\sin \pi x}$$

has a denominator with zeros $0, \pm 1, \pm 2, \ldots$, and should therefore be expressible in the form

$$\cot \pi x = \frac{a}{x} + \sum_{n=1}^{\infty} \left(\frac{b_n}{x-n} + \frac{c_n}{x+n} \right). \tag{7}$$

From this, the constants a, b_n, and c_n can be found by the procedure suggested (they are all equal to $1/\pi$), and (7) can then be rearranged to yield (6). For reasons that will now be obvious, it is customary to refer to (6) as the *partial fractions expansion of the cotangent*. The main gap in this suggestive but rather tentative derivation is of course the fact that we have no prior guarantee that an expansion of the form (7) is possible.

Another way of approaching (6) is to begin with the infinite product (6) in Section B.14:

$$\frac{\sin x}{x} = \left(1 - \frac{x^2}{\pi^2}\right)\left(1 - \frac{x^2}{4\pi^2}\right)\left(1 - \frac{x^2}{9\pi^2}\right)\cdots\left(1 - \frac{x^2}{n^2\pi^2}\right)\cdots.$$

If we take the logarithm of both sides to obtain

$$\ln \frac{\sin x}{x} = \sum_{n=1}^{\infty} \ln \left(1 - \frac{x^2}{n^2\pi^2}\right),$$

and then differentiate, the result is seen to be

$$\cot x - \frac{1}{x} = \sum_{n=1}^{\infty} \frac{-2x}{n^2\pi^2 - x^2}$$

or

$$\cot x = \frac{1}{x} + 2x \sum_{n=1}^{\infty} \frac{1}{x^2 - n^2\pi^2};$$

and replacing x by πx and then multiplying through by πx yields

$$\pi x \cot \pi x = 1 + 2x^2 \sum_{n=1}^{\infty} \frac{1}{x^2 - n^2}, \tag{8}$$

which is equivalent to (6).

Euler's Formula for $\sum \frac{1}{n^{2k}}$

We now obtain a major payoff from (5) and (8) by replacing x by πx in (5) and equating the two expressions for $\pi x \cot \pi x$,

$$1+\sum_{n=1}^{\infty} \frac{-2x^2}{n^2-x^2} = 1+\sum_{k=1}^{\infty}(-1)^k \frac{2^{2k} B_{2k}}{(2k)!}(\pi x)^{2k}, \tag{9}$$

where we use k as the index of summation on the right for reasons that will appear in a moment. Each term of the series on the left is easy to expand in a geometric series,

$$\frac{-2x^2}{n^2-x^2} = -2\frac{x^2/n^2}{1-x^2/n^2} = -2\sum_{k=1}^{\infty}\left(\frac{x^2}{n^2}\right)^k = -2\sum_{k=1}^{\infty}\frac{x^{2k}}{n^{2k}},$$

so (9) can be written as

$$1+\sum_{n=1}^{\infty}\left(-2\sum_{k=1}^{\infty}\frac{x^{2k}}{n^{2k}}\right) = 1+\sum_{k=1}^{\infty}(-1)^k \frac{2^{2k} B_{2k}}{(2k)!}\pi^{2k} x^{2k}.$$

We now interchange the order of summation on the left and obtain

$$1+\sum_{k=1}^{\infty}\left(-2\sum_{n=1}^{\infty}\frac{1}{n^{2k}}\right) x^{2k} = 1+\sum_{k=1}^{\infty}(-1)^k \frac{2^{2k} B_{2k}}{(2k)!}\pi^{2k} x^{2k},$$

and equating the coefficients of x^{2k} yields

$$\sum_{n=1}^{\infty}\frac{1}{n^{2k}} = (-1)^{k-1}\frac{2^{2k} B_{2k}}{2(2k)!}\pi^{2k}$$

for each positive integer k. In particular, for $k = 1, 2, 3$ we get

$$\sum_{n=1}^{\infty}\frac{1}{n^2} = \frac{\pi^2}{6}, \qquad \sum_{n=1}^{\infty}\frac{1}{n^4} = \frac{\pi^4}{90}, \qquad \sum_{n=1}^{\infty}\frac{1}{n^6} = \frac{\pi^6}{945}.$$

It is very remarkable that for almost 250 years there has been no progress whatever toward finding the exact sum of any one of the series

$$\sum_{n=1}^{\infty}\frac{1}{n^3}, \qquad \sum_{n=1}^{\infty}\frac{1}{n^5}, \qquad \sum_{n=1}^{\infty}\frac{1}{n^7}, \ldots$$

Perhaps a second Euler is needed for this breakthrough, but none is in sight.

Annotated Bibliography

[1] M. E. Baron, *The Origins of the Infinitesmal Calculus*, Pergamon Press, Oxford, 1969.

In the Preface, the author states "the origins of the infinitesmal calculus lie, not only in the significant contributions of Newton and Leibniz, but also in the centuries-long struggle to investigate area, volume, tangent and arc by purely geometric means." She decided to emphasize the geometric techniques and methods employed first by the Greek, Hindu, and Arabic cultures, and then by the European predecessors of Newton and Leibniz. Her early study was enhanced by Boyer's book (see [**4**]) including its extensive references. This book is thorough and well-referenced. Its scholarly development makes it more suitable for faculty than students.

[2] Eric Temple Bell, *Men of Mathematics*, Simon and Schuster, New York, 1961.

This book was written in 1935 at a time when expository books about mathematics and mathematicians were not readily available. It came as a breath of fresh air with its lively accounts of biographical details and mathematical accomplishments for many of the greatest mathematicians of history. Ever since, it has inspired many to learn more about mathematics. While some have criticized the author for factual inaccuracies he includes to enliven the exposition, it remains a valuable resource. Compare the nature of the writing in this book with the more formal exposition in the Biographical Dictionary of Mathematicians (see [**16**]).

[3] Garrett Birkhoff, *A Source Book in Classical Analysis*, Harvard University Press, Cambridge, MA, 1973.

This source book of mathematical writings in the nineteenth century picks up where Struik's source book (see [**31**]) ends. The writings that are most helpful for a course in real analysis are in Chapters 1, 3, and 5. They include material from Cauchy, Fourier, Gauss, Abel, Bolzano, Riemann and Weierstrass.

[4] C. B. Boyer, *The History of the Calculus and its Conceptual Development*, Dover, New York, 1959.

This book was written in 1949 to provide a historical treatment of the "fundamental ideas of the subject (i.e. calculus) from their incipiency in antiquity to the final formulation of these in the precise concepts familiar to every student of the elements of modern mathematical analysis." This quote is taken from the Preface. As a 300 page paperback, it provides a readable and helpful introduction to this topic. A substantial bibliography suggests additional resources. The author also wrote a complete history of mathematics text in 1968[1] which is distinguished by the inclusion of an extensive number of individuals and results, but each section is quite brief (about one page in length on average).

[5] David Bressoud, *A Radical Approach to Real Analysis*, The Mathematical Association of America, 1994.

The Preface explains this book "is designed to be a first encounter with real analysis, laying out its context and motivation in terms of the transition from power series to those that are less predictable, especially Fourier series, and marking some of the traps into which even great mathematicians have fallen. The book begins with Fourier's introduction of trigonometric series and the problems they created for the mathematicians of the early nineteenth century. It follows Cauchy's attempts to establish a firm foundation for calculus, and considers his failures as well as his successes. It culminates with Dirichlet's proof of the validity of the Fourier series expansion and explores some of the counterintuitive results Riemann and Weierstrass were led to as a result of Dirichlet's proof."

This is a radical approach as contrasted with a traditional approach of organizing the analysis course in the logical order of concept development. The order of topics in such a course is the real number properties, sequences, continuity, differentiability, integrability, and sequences and series of functions, without regard for the historical development of analysis. The main focus of this book is on nineteenth century material, and especially trigonometric series. Many graphical and numerical investigations are provided to encourage active exploration by the reader.

[6] David M. Burton, *History of Mathematics: An Introduction*, Third edition, Wm. C. Brown Publishers, Dubuque, IA, 1995.

A distinguishing characteristic of Burton's book is its emphasis on well-written prose. Some mathematical details are omitted in favor of providing interesting narrative of the topics he chooses to present. There are separate chapters devoted to Fibonacci sequences, the cubic and quartic equations, probability theory, and non-Euclidean geometry, while algebra and analysis receive slight attention. The material he does present on Newton and Leibniz in Chapter 8 is very good, but the material on Cauchy, Riemann, and Weierstrass in Section 11.3 is rather meager.

---

[1] Carl B. Boyer and Uta C. Merzbach, *A History of Mathematics*, Second edition, John Wiley and Sons, New York, 1989.

[7] Ronald Calinger, editor, *Classics of Mathematics*, Prentice Hall, Inc, 1995.

This is a collection of excerpts from the original writings of several mathematicians that have been translated into English. There are also brief biographical selections and comments about the importance of the material. Some samples that are pertinent to analysis include the following:

(a) Gottfried Leibniz and the Fundamental Theorem of Calculus on pages 383–386 and 393–394.

(b) Colin Maclaurin and the *Treatise of Fluxions* on pages 475–478.

(c) Niels Abel and the binomial series on pages 537–538 and 594–596.

(d) Augustin-Louis Cauchy and the *Cours d'analyse* on pages 597–603.

(e) Karl Weierstrass and the differential calculus on pages 604–610.

[8] John H. Conway and Richard K. Guy, *The Book of Numbers*, Springer-Verlag, New York, 1996.

This interesting book is a collection of short descriptions about unusual properties of familiar numbers. While most fit best in a course on discrete mathematics, number theory, algebra or geometry, there are several items of interest for analysis. These include the use of Gregory's and Stormer's numbers for the approximation of π, the Euler–Mascheroni[2] number (γ), Apery's number ($\zeta(3)$), Bernoulli numbers resulting from Faulhaber's formula, and various methods to find the pattern in a sequence of integers.

[9] William Dunham, "A Historical Gem from Vito Volterra" *Mathematics Magazine*, 63 (1990) 234–237.

Many examples that were considered "unusual" or even "pathological" at the time of their introduction have often led to an improved understanding of basic analysis concepts as continuity, differentiability, and integration. Many of these examples such as Dirichlet's everywhere discontinuous function (see Problem 7) are characterized by the use of one expression for rational domain values and a different expression for irrational domain values. This article deals with one such function that is continuous only for irrational values, using a proof by Volterra in the 1880s to demonstrate the impossibility of finding a function that is continuous only for rational values.

[10] William Dunham, *Euler: The Master of Us All*, The Mathematical Association of America, Washington, D.C., 1999.

This paperback is Volume 22 in the MAA's Dolciani Mathematical Expositions series. It begins with a short biographical sketch of Euler and contains eight chapters on topics where Euler significantly advanced the level of knowledge. In Chapter 2 on logarithms, Euler describes the early usage of series expansions to represent functions as $\ln(1+x)$ and e^x without the usual approach of derivatives and Maclaurin series. Chapter 3 on infinite series traces Euler's attempts over a

[2] Lorenzo Mascheroni calculated the value for Euler's constant to 32 decimal places in 1790, but there were errors after the twentieth decimal place.

period of years to find the exact value for

$$\sum_{n=1}^{\infty} \frac{1}{n^2}.$$

Chapter 4 presents some connections between infinite series and the primes of number theory, and contains various proofs that the series of reciprocals of the primes diverges to infinity. Chapter 5 introduces the imaginary numbers and has three proofs of Euler's identity that $e^{ix} = \cos x + i \sin x$.

[11] William Dunham, *Journey Through Genius*, Penguin Books, New York, 1991.

This little book is popular with students because of its readability and interesting historical and biographical details. Four of the chapters are appropriate for an analysis course. Chapter 4 portrays the Greek period in the time of Archimedes and his idea of approximating areas and volumes by some very distinctive methods. Chapter 7 presents some of the gifted individuals in the 17th century who preceded Isaac Newton, and includes his work with the general binomial theorem and its use to approximate π. Chapter 8 contains the work of Leibniz and the Bernoullis to provide an approach to calculus that differed from Newton's, and includes an unusual proof of the divergence of the harmonic series. Chapter 9 presents the prolific life and accomplishments of Leonhard Euler and his discovery of the formula

$$\sum_{n=1}^{\infty} \frac{1}{n^2} = \frac{\pi^2}{6}.$$

The author is writing a similar book presenting some important results from analysis which is to be published by Princeton University Press. This new book will be a valuable addition to this list of references.

[12] C. H. Edwards, Jr., *The Historical Development of the Calculus*, Springer-Verlag, New York, 1979.

This book presents the historical development of calculus throughout the entire period of mathematical history. More than half of the book, seven chapters, deals with the contribution of individuals prior to the time of Newton and Leibniz. There is a chapter each on Newton and Leibniz, and three chapters on the subsequent development of this subject. The author does a nice job in presenting each result in terms of the approach that was originally used. Emphasis is on the methods used to solve calculus problems throughout history, and many exercises are included.

[13] Howard Eves, *An Introduction to the History of Mathematics*, 6th ed., Saunders College Publishing, 1990.

This history text was written in 1964 with the goal of presenting the history of mathematics to prospective secondary teachers of mathematics. It is distinguished by extensive problem sets at the end of every chapter. Among other additions, the sixth edition includes ten cultural connections written by Jamie Eves. The material on calculus is mainly in Chapters 11 and 12, although there is a very nice chronology of π in Chapter 4.

[14] John Fauvel and Jeremy Gray, editors, *The History of Mathematics: A Reader*, Macmillan Press, London, 1987.

This collection of readings was assembled to support a course on the history of mathematics that was offered at the Open University in England. Its value for students is enhanced because the selections have been translated into English when necessary, and because of many unusual items such as obituary notices and excerpts from letters. For example, the reader can find a description of the first meeting between John Bernoulli and the Marquis de l'Hospital which can enliven a calculus presentation of l'Hospital's rule (see Selection 1 in Part IV). Most of the entries that pertain to calculus and analysis are found in Chapters 12–14 and 18.

[15] A. Gardiner, *Infinite Processes: Background to Analysis*, Springer-Verlag, New York, 1982.

The purpose of this book is to carefully examine the infinite processes that arise in elementary mathematics as a prologue to analysis. The longest unit is Part II which rigorously develops the concept of number. Part III develops the concept of geometry, and the status of number in this setting. The author points out that geometry needed to be set aside for the careful presentation of calculus around 1870, based on an arithmetic treatment of real numbers. Part IV discusses the question: What is a function?

[16] Charles Coulston Gillispie, Editor-in-chief, *Biographical Dictionary of Mathematicians*, Charles Scribner's Sons, New York, 1995.

This is a four volume set that provides detailed biographical sketches of all important mathematicians, along with some commentary about the content and significance of their work. A copy should be in every mathematics department library. The material on mathematicians is also contained in the 17 volume set titled the *Scientific Dictionary of Biography*.

[17] Judith Grabiner, "Is Mathematical Truth Time Dependent?" *The American Mathematical Monthly*, 81 (1974) 354–365.

Judith Grabiner has written articles and books and given lectures about some of the foundational issues in calculus. The AMS-MAA Joint Lecture Series has a video of a talk she delivered at the January 1991 annual meetings in San Francisco with the title "Was Newton's Calculus Just a Dead End: Maclaurin and the Scottish Connection."

In this article, she looks at the changing level of rigor in calculus from the time of Newton and Leibniz at the end of the 17th century until the time of Cauchy and Weierstrass in the 19th century, and asks for the reasons of this revolution in thought. After giving some examples of the reliance by the 18th century workers in calculus on any method that gives new results, she lists several reasons why standards changed so dramatically in the 19th century to produce a much higher level of rigor. Individuals such as Cauchy, Riemann, and Weierstrass took 18th century approximation techniques and transformed them into the clear definitions and rigorous proofs that characterized 19th century in calculus. This article is included as Selection 3 in Part IV.

[18] Judith Grabiner, *The Origins of Cauchy's Rigorous Calculus*, The MIT Press, Cambridge, MA, 1981.

Augustine-Louis Cauchy was largely responsible for the initiation of rigor into calculus. The author develops the position that Cauchy's crucial insight was that "by means of the limit concept, the calculus could be reduced to the algebra of inequalities." She also shows in this book that while all parts of calculus received a new formulation at the hands of Cauchy, many predecessors contributed to this and many successors carried out his work. The contributions of Euler and Lagrange during the 18th century are highlighted. Gauss and Bolzano are presented as two individuals who had similar ideas as Cauchy but who were distracted by other pursuits. Riemann and Weierstrass are later contemporaries of Cauchy who pushed his ideas further. The author devotes the second half of the book to a presentation of Cauchy's results on the calculus, showing how it "made obsolete many of the earlier treatments of limits, convergence, continuity, derivatives, and integrals." A helpful appendix contains English translations of some of Cauchy's major contributions to the foundations of the calculus.

[19] Ivor Grattan-Guinness, *The Development of the Foundations of Mathematical Analysis from Euler to Riemann*, The MIT Press, Cambridge, MA, 1970.

This book opens with a discussion of the vibrating string problem as it was understood in the 18th century by d'Alembert and Euler. Work on this question continued during the next century, leading to the method of trigonometric series as developed by Daniel Bernoulli, Lagrange, and then Joseph Fourier, who used this device for his solution of the heat diffusion equation. Although there is some discussion of more general ideas from the development of analysis in the 19th century by Cauchy, Dirichlet, Riemann, and Weierstrass, the dominant theme is that of Fourier series and convergence questions related to it. Many examples are presented in careful detail to illustrate how various questions were handled in their historical context. The book ends with Riemann's integrability criterion, and the author notes that this led to a new era of analysis in which the ideas of Cantor's set theory and Lesbegue's measure theory dominate. There is also an appendix which presents the various convergence tests for series of constants that were developed during the first half of the 19th century.

[20] Ivor Grattan-Guinness, editor, *From the Calculus to Set Theory 1630–1910*, Princeton University Press, 2000.

This book was first published by Gerald Duckworth and Co. in London in 1980, but was out of print before that decade ended. Fortunately, a paperback version is now available. The book contains six chapters, each written by a different author on a topic of their speciality. In the first chapter, Kristi Moller Pederson presents the beginning of the ideas of calculus during the period 1630 to 1660, while in Chapter 2 Henk Bos continues this development from Newton and Leibniz until the middle of the 18th century. I. Grattan-Guinness writes Chapter 3 to trace the emergence of analysis ideas from physical problems in the 18th century through the search for foundations during most of the 19th century. Thomas Hawkins discusses in Chapter 4 the development of the integral concept that culminated with Lebesgue measure and integration in 1902. Joe Dauben presents the basic results of Cantor's set theory in Chapter 5, and Robert Bunn writes the concluding chapter on the foundations of mathematics, especially in logic, that had developed by the time Russell and Whitehead wrote *Principia mathematica* around 1910. The first four chapters of the book provide an excellent historical framework for the development of analysis, along with several specific examples from the mathematics of this time period.

[21] E. Hairer and G. Wanner, *Analysis by Its History*, Springer-Verlag, New York, 1995.

The authors choose to treat the topics of analysis in their historical order. Chapter 1 deals with the origins of the elementary functions using the mathematical reasoning of Descartes, Newton, and Euler. Chapter 2 introduces the differential and integral calculus in the language of Leibniz, the Bernoullis, and Euler. Chapter 3 presents the theoretical foundations of calculus from the nineteenth century, correcting errors and replacing intuitive reasoning of the previous two centuries. The final chapter treats calculus of several variables. An abundance of interesting quotations enliven the book. A wide range of major theorems are included, well beyond the scope of a usual one semester course in real analysis.

[22] Thomas Hawkins, *Lebesgue's Theory of Integration: Its Origins and Development*, Chelsea Press, New York, 1975.

The first two chapters present Riemann's development of the concepts of function and integral in the 1850s as a significant improvement over the work of his predecessors such as Cauchy. The next two chapters discuss Cantor's theory of sets and the subsequent development of ways to measure these sets during the period from 1870 to 1900. The pioneering results of Lebesgue to create a modern theory of integration in the opening decades of the twentieth century are given in Chapter 5. The author carefully places the new theory in the setting of its historical context and development as a reasonable consequence of the work of many individuals in the fifty years preceding the year 1902.

[23] Omar Hijab, *Introduction to Calculus and Classical Analysis*, Springer-Verlag, New York, 1997.

This book provides a non-traditional choice for a real analysis text. The author chooses to include many significant applications and devotes a quarter of the book to this topic. Some of the more unusual inclusions for a book at this level are the Euler–Maclaurin formula, the Riemann zeta function, continued fractions, infinite products, and Laplace and Fourier transforms. On the pedagogical side, he chooses to avoid ϵ-δ arguments, replacing them by sequence results based, in turn, on the sup and inf concepts. The integral is then defined in terms of area under the graph, and the interchange of limits and integral theorems are based on the monotone convergence theorem. In the Preface, the author states that he "chose several of the jewels of classical eighteenth and nineteenth century analysis and inserted them at the end of the book, inserted the axioms for reals at the beginning, and filled in the middle with (and only with) the material necessary for clarity and logical completeness." Each teacher will need to decide whether this book is suitable as a text for his or her students, or if it works better as a reference source for ideas.

[24] Dan Kalman, "Six Ways to Sum a Series," *The College Mathematics Journal*, Vol. 24, No. 5, 1993, pp. 402–421.

This interesting article delivers just what its title promises. The series referred to is Euler's series, namely

$$\sum_{n=1}^{\infty} \frac{1}{n^2}$$

which converges to $\pi^2/6$. A rich collection of mathematical results and methods are presented, including unusual trigonometric identities, de Moivre's formula, the tranformation of a double integral using Jacobians, residue theory from complex analysis, Fourier series, and vector spaces. Most of these results are adapted from other sources. Additional proofs that

$$\sum_{n=1}^{\infty} \frac{1}{n^2} = \frac{\pi^2}{6}$$

continue to regularly occur in the literature. It seems a shame there are so many different proofs for the value of

$$\sum_{n=1}^{\infty} \frac{1}{n^2},$$

but not one for the exact value of

$$\sum_{n=1}^{\infty} \frac{1}{n^3}.$$

[25] Victor J. Katz, *A History of Mathematics*, Addison-Wesley Educational Publishers, Inc., New York, 1998.

Katz's book was the first history of mathematics text to give serious attention (i.e., three chapters) to the contribution from cultures other than those of the early Greek and the modern European periods that are traditionally emphasized. His work is also distinguished by his attention to mathematical details that support the concepts presented in the historical narrative. Three chapters are devoted to calculus and analysis for a total of 176 pages.

[26] Morris Kline, *Mathematical Thought from Ancient to Modern Times*, Oxford University Press, New York, 1972.

This lengthy book has an entry on most major topics in the development of mathematics. Five of the fifty chapters are devoted to calculus and real analysis, beginning with Chapter 17 titled "The Creation of the Calculus" and ending with Chapter 40 titled "The Instillation of Rigor in Analysis." Other chapters deal with parts of analysis such as four chapters on differential equations, two chapters on the calculus of variations, and single chapters on functions of a complex variable, theory of functions of a real variable, functional analysis, and divergent series. Kline is able to express the main ideas in a readable manner.

[27] Reinhard Laubenbacher and David Pengelley, *Mathematical Expeditions*, Springer-Verlag, New York, 1999.

This book is authored by two members of the mathematics faculty at New Mexico State University, where there has been a serious attempt to use original sources at many levels of mathematics instruction. This book gives a historical introduction to five main areas of mathematics, followed by some representative selections of original work by important figures in the development of each topic. In Chapter 3 on analysis, the selections are chosen from the writings of Archimedes, Cavalieri, Leibniz, Cauchy, and Abraham Robinson.

[28] George F. Simmons, *Calculus Gems*, McGraw Hill, New York, 1992.

This small paperback was originally an appendix to Simmons' calculus text in the 1980s. Part 1 contains brief biographies of several individuals who contributed in some way to the development of the ideas of calculus. These are uneven in that some of the most important individuals, such as the Bernoullis, Fourier, and Cauchy have very brief selections, while some minor individuals have extended entries. Part 2 contains several "gems," which are outlines of proofs for some interesting results not typically found in a calculus text. Some of the more interesting ones include the catenary curve, the Bernoulli numbers, the general Maclaurin series for $\tan x$, and several proofs that

$$\sum_{n=1}^{\infty} \frac{1}{n^2} = \frac{\pi^2}{6}.$$

See Selection 5 in Part IV for one of these gems.

[29] Saul Stahl, *Real Analysis: A Historical Approach*, John Wiley & Sons, New York, 1999.

The first five chapters contain examples of the work of Archimedes, Fermat, Newton and Euler on topics from infinite series, power series, and trigonometric series. The middle of the book develops the basic properties of real numbers, sequences, continuity, differentiability, and infinite series. Material on the Riemann integral is omitted to make room for material on the historical movitation for the concepts of convergence. Many interesting problems are presented, including a significant amount of material on Fourier series.

[30] John Stillwell. *Mathematics and Its History*, Springer-Verlag, New York, 1989.

Chapter 8 briefly describes several of the discoveries in the 17th century by individuals such as Cavalieri, Fermat, and Wallis leading up to the calculus as formulated by Newton and Leibniz. Chapter 9 on infinite series contains material about interpolation formulas, generating functions, and the zeta function. Each of the twenty chapters concludes with short biographical sketches of mathematicians who were influential in the development of the mathematics discussed in that chapter.

[31] Dirk J. Struik, *A Source Book in Mathematics 1200–1800*, Harvard University Press, Cambridge, MA, 1969.

This book contains five chapters with selections of mathematical writings which have been translated into English. Most selections are brief, averaging about six pages, with a short introduction for each one. Chapters 4 and 5 contain 38 selections dealing with topics from analysis, written by individuals such as Kepler, Cavalieri, Fermat, Pascal, Leibniz, Newton, Euler, and the Bernoullis. This author has also written *A Concise History of Mathematics*, published in a Dover paperback in 1987.

[32] Robert M. Young, *Excursions in Calculus: An Interplay of the Continuous and Discrete*, The Mathematical Association of America, 1992.

The claim is made that "the purpose of this book is to explore—within the context of elementary calculus—the rich and elegant interplay that exists between the two main currents of mathematics, the continuous and the discrete." As might be expected, there are many results from number theory and the search for patterns, dealing with their implications for and applications to many topics in the calculus. There is a collection of more than 400 exercises, with a reference provided for the solution of each one. More than 400 references are also included for additional reading and research.

[33] Hans Niels Jahnke, editor, *A History of Analysis*, American Mathematical Society, 2003.

I include this reference at the end of the Annotated Bibliography since I only became aware of it shortly before my book went to press. It is a joint effort of several individuals who each wrote a chapter on part of the history of analysis from antiquity until the end of the nineteenth century. There was much interaction between the writers to ensure a unified text, most of which was translated from the original German into English. It is an excellent, detailed, scholarly treatment which will be of great benefit to the instructor.

Additional References

Several references to helpful books and articles that are not listed in the Annotated Bibliography occur throughout the text. These are listed below and the number in parentheses following each entry refers to the page in the text where that reference may be found.

Donald J. Albers, G. L. Alexanderson, Constance Reid, *International Mathematical Congresses: An Illustrated History*, Springer-Verlag, New York, 1987. (page 142)

E. N. da C. Andrade, *A Brief History of the Royal Society*, Royal Society, London, 1960. (page 129)

Robert G. Bartle, *The Elements of Integration*, John Wiley & Sons, Inc., New York, 1966. (page 122)

Bruno Belhoste, *Augustin-Louis Cauchy: A Biography*, Springer-Verlag, New York, 1991. (page 135)

Bruce C. Berndt, *Ramanujan's Notebooks*, Part I, Springer-Verlag, New York, 1985. (page 145)

Carl B. Boyer and Uta C. Merzbach, *A History of Mathematics*, Second edition, John Wiley & Sons, New York, 1989. (pages 139 and 144)

Robert L. Brabenec, *Introduction to Real Analysis*, PWS-Kent Publishing Company, Boston, 1990. (page 126)

R. Creighton Buck, *Advanced Calculus*, McGraw-Hill Book Company, New York, 1965. (page 107)

W. K. Bühler, *Gauss: A Biographical Study*, Springer-Verlag, New York, 1981. (page 138)

Gale E. Christianson, *In the Presence of the Creator: Isaac Newton and His Times*, The Free Press, New York, 1984. (page 129)

Joseph Warren Dauben, *Georg Cantor: His Mathematics and Philosophy of the Infinite*, Harvard University Press, Cambridge, Massachusetts, 1979. (page 142)

John Derbyshire, *Prime Obsession: Bernhard Riemann and the Greatest Unsolved Problem in Mathematics*, Joseph Henry Press, Washington D.C., 2003. (page 147)

Marcus Du Sautoy, *The Music of the Primes*, HarperCollins Publisher, New York, 2003. (page 147)

Underwood Dudley, *Elementary Number Theory*, Second edition, W. H. Freeman and Company, San Francisco, 1978. (page 169)

William B. Ewald, ed., *From Kant to Hilbert: A Source Book in the Foundations of Mathematics*, Clarendon Press, Oxford, 1996. (page 140)

Bernard R. Gelbaum and John M. H. Olmsted, *Counterexamples in Analysis*, Holden-Day, Inc., San Francisco, 1964. (page 36)

I. Grattan-Guinness, ed., *Companion Encyclopedia of the History and Philosophy of the Mathematical Sciences*, Routledge Inc., New York, 1994. (page 170)

Jeremy J. Gray, *The Hilbert Challenge*, Oxford University Press, Oxford, 2000. (page 144)

Thomas L. Hankins, *Jean d'Alembert: Science and the Enlightenment*, Clarendon Press, Oxford, 1970. (page 134)

G. H. Hardy, *A Mathematician's Apology*, Cambridge University Press, Cambridge, 1969. (page 145)

Julian Havil, *Gamma: Exploring Euler's Constant*, Princeton University Press, Princeton, New Jersey, 2003. (pages 33 and 146)

John Herival, *Joseph Fourier: The Man and the Physicist*, Clarendon Press, Oxford, 1975. (page 136)

Joseph E. Hofmann, *Leibniz in Paris 1672–1676: His Growth to Mathematical Maturity*, Cambridge University Press, Cambridge, 1974. (page 130)

Ioan James, *Remarkable Mathematicians*, Cambridge University Press, Cambridge, 2002. (page 144)

Robert Kanigel, *The Man Who Knew Infinity*, Charles Scribner's Sons, New York, 1991. (page 145)

Don H. Kennedy, *Little Sparrow: A Portrait of Sophia Kovalevsky*, Ohio University Press, Athens, Ohio, 1983. (page 141)

James Kirkwood, *An Introduction to Analysis*, PWS-Kent Publishing Company, Boston, 1989. (page 43)

Ann Hibner Koblitz, *A Convergence of Lives Sofia Kovalevskaia: Scientist, Writer, Revolutionary*, Rutgers University Press, New Brunswick, New Jersey, 1993. (page 141)

Eli Maor, *e: The Story of a Number*, Princeton University Press, Princeton, New Jersey, 1994. (page 82)

Michael Monastyrsky, *Riemann, Topology, and Physics*, translated by James King and Victoria King and edited by R. O. Wells, Jr., Birkhäuser, Boston, 1999. (page 140)

James R. Newman, editor, *The World of Mathematics*, Simon and Schuster, New York, 1956. (page 143)

Constance Reid, *Hilbert*, Springer-Verlag, New York, 1970. (page 142)

Ranjan Roy, "The Discovery of the Series Formula for π by Leibniz, Gregory, and Nilakantha," *Mathematics Magazine*, Vol. 63, No. 5, December 1990, pp. 291–306. (page 127)

George Sarton, "Lagrange's Personality," *Proceedings of the American Philosophical Society*, Vol. 88, No. 6, 1944. (page 134)

James A. Sellers, "Beyond Mere Convergence," PRIMUS, Volume XII, Number 2, June 2002, pp. 157–164. (page 94)

N. J. A. Sloane, *The On-Line Encyclopedia of Integer Sequences*, Published electronically as http://www.research.att.com/~njas/sequences. (page 95)

Ian Tweddle, *James Stirling*, Scottish Academic Press, Edinburgh, 1987. (pages 179–187)

Alfred van der Poorten, "A Proof that Euler Missed," *The Mathematical Intelligencer*, Vol. 1, No. 4, 1978, pp. 195–203. (page 82)

David V. Widder, *Advanced Calculus*, 2nd ed., Prentice-Hall, Inc., Englewood Cliffs, New Jersey, 1961. (page 119)

Raymond L. Wilder, *Introduction to the Foundations of Mathematics*, John Wiley & Sons, New York, 1965. (page 166)

Benjamin H. Yandell, *The Honors Class: Hilbert's Problems and Their Solvers*, A K Peters, Natick, MA, 2002. (page 142)

Index

About the Author

Robert L. Brabenec was born on January 11, 1939 in Chicago, Illinois. His grandparents came to America from Poland and Czechoslovakia. He majored in mathematics and received a BS with highest honors from Wheaton College in 1960, and a PhD in mathematics from Ohio State University in 1964, with a dissertation on the topic of measure and integration theory in a linear vector space. He began teaching mathematics at Wheaton College in June 1964, and after a two-year tour of active duty as an Army captain in missile intelligence, he returned to Wheaton in 1967 to become chair of the newly-formed Department of Mathematics, and he has continued in that role for thirty-six years. He received the Teacher of the Year award from Wheaton College in 1970. He has also taught part-time at the University of Alabama at Huntsville and at Northwestern University.

He has been a member of the Mathematical Association of America since 1960 and a long-term member of the American Mathematical Society as well. In 1977, he began the organization that is now called the Association of Christians in the Mathematical Sciences and has continued to serve as the Executive Secretary until the present. This organization of about 400 members sponsors a major conference every two years, and supports other activities to help mathematicians see connections between mathematics and religious faith.

Over the years, he has developed a strong interest in the history and foundations of mathematics, integrating these ideas into most courses that he teaches. He is author of a textbook *Introduction to Real Analysis* that which was published by PWS-Kent in 1990, and has presented several articles, talks, and a MAA minicourse dealing with historical themes in mathematics. He spent sabbaticals in 1988 at the Universities of London and Cambridge, in 1995 at the University of Virginia, and in 2002 back at Cambridge University. In a typical year, his teaching load consists of a calculus course, the mathematics colloquium course, and the required major courses of real analysis, as well as the history and foundations of mathematics.